Reliability and Maintenance - an Overview of Cases

Edited by Leo Kounis

Published in London, United Kingdom

IntechOpen

Supporting open minds since 2005

Reliability and Maintenance - an Overview of Cases
http://dx.doi.org/10.5772/intechopen.77493
Edited by Leo Kounis

Contributors
Yan Ran, Zongyi Mu, Wei Zhang, Genbao Zhang, Qian Wang, Abdullah Mohammed Al-Shaalan, Razzaqul Ahshan, Vanderley Vasconcelos, Antônio C. L. Costa, Amanda L. Raso, Wellington A. Soares, Franco Robledo, Luis Stábile, Pablo Romero, Omar Viera, Pablo Sartor, Wu Zhou, Jiangbo He, Carmen Patino-Rodriguez, Dr. Fernando Guevara, Nam-Ho Kim, Raphael T. Haftka, Ting Dong

First published in London, United Kingdom, 2020 by IntechOpen
IntechOpen is the global imprint of INTECHOPEN LIMITED, registered in England and Wales, registration number: 11086078, 7th floor, 10 Lower Thames Street, London, EC3R 6AF, United Kingdom
Printed in Croatia

British Library Cataloguing-in-Publication Data
A catalogue record for this book is available from the British Library

Additional hard and PDF copies can be obtained from orders@intechopen.com

Reliability and Maintenance - an Overview of Cases
Edited by Leo Kounis
p. cm.
Print ISBN 978-1-78923-951-5
Online ISBN 978-1-78923-952-2
eBook (PDF) ISBN 978-1-83880-736-8

For EU product safety concerns:
IN TECH d.o.o., Prolaz Marije Krucifikse Kozulić 3, 51000 Rijeka, Croatia, info@intechopen.com or visit our website at intechopen.com.

Meet the editor

Leo D. Kounis is Head of Department of the Communication and Informatics Battalion of the Hellenic Ministry of Defense, Hellenic National Defense General Staff. He obtained his BEng (Hons) degree in Manufacturing Systems Engineering, his MSc in Quality Engineering, and his PhD in Systems Reliability from the University of Hertfordshire, UK. Dr. Kounis has worked as a senior quality engineer in a number of private companies in Greece and has acted as part-time lecturer and scientific advisor in academia. His research interests focus in the area of quality, transportation, and sustainable energy. He has published a number of scientific papers.

Contents

Preface

In today's highly automated and digitalized world, the terms *reliability, maintenance*, and *availability* (RMA) are considered as forming the core part of the contemporary reliability engineering discipline. Indeed, system engineers and logisticians may be regarded as being the primary users of the methods and techniques in RMA. However, because of continuous automation and acknowledging the fact that computers and microchips find their way into countless modern-day applications, products, and systems, which in turn require increased levels of RMA, the latter impact their associated lifecycle costs and their usefulness. Hence, reliability techniques are equally applied to hardware and software.

Thus, the term *reliability*, although having a number of definitions, is generally accepted to refer to the degree to which a system, product, or component performs its intended functions under stated conditions for a specified period of time without failure. In addition to the above definition, the terms *maintainability* and *availability* are also used to describe in the former case the probability of a system, product, or component to be repaired in a defined environment within a specified time period and that the repaired system, product, or component in the latter instance is readily operational/functional. The American Society for Quality, as well as the International Standards Organization, among other renowned bodies provide detailed and area-specific definitions.

Amid a plethora of challenges, the most important one being ongoing climate change, systems engineering areas will be called upon to deliver products and offer services to boost a higher degree of reliability, maintainability, and availability, placing an emphasis on designing for reliability and postproduction management systems.

This book comprises nine chapters split in two sections.

The first section discusses maintenance models and policies.

The first chapter introduces a contemporary maintenance strategy in line with the ISO 55000 series of asset management standards. The latter may be regarded as the successor to Publicly Available Specification PAS55 of the British Standards Institution. The suggested strategy was validated in electric power generation systems and transport vehicles.

The advantages of condition-based maintenance over scheduled maintenance regarding the safety and lifetime cost of an aircraft fuselage are discussed in the second chapter.

Meta-action units as the basic analysis and control unit concerning computer numerical-controlled machines are presented in the next chapter, which also portrays an overview of the respective reliability technology.

A number of surrogate modeling methods have been introduced to reduce model-specific evaluation time and are applied in cases in which the outcome of interest may not be measured in a straightforward and unequivocal manner. The following research work discusses a method based on radial basis functions aimed at probabilistic analysis applications.

The last chapter of the first section focuses on the basic mechanisms pertaining to specific reliability issues, such as thermal drift and long-term storage drift observed in microelectromechanical systems, by providing the corresponding reliability analysis of the performed experiments.

The second section presents four chapters on reliability issues in industrial networks.

The opening chapter focuses on the resolution of a mixed model regarding the design of large-size networks, by introducing an algorithm concerning connectivity and reliability by combining network survivability and network reliability approaches.

The treatment of uncertainties in probabilistic risk assessment is the subject area of the next chapter, which investigates uncertainty handling approaches pertaining to the analysis of fault tress and event trees as a means to overcome observed ambiguities.

A number of fundamental concepts concerning the reliability evaluation of power systems are discussed in this chapter by deriving a number of measures, criteria, and performance-related indices.

In a similar case, the last chapter discusses the reliability evaluation of a microgrid system acknowledging the intermittency effect of renewable energy sources, such as wind, by utilizing the Monte Carlo simulation technique.

It is hoped that the outcomes presented herein may serve as a platform for further research.

Dr. Leo D. Kounis
Hellenic Ministry of Defense,
Communication and Informatics Battalion,
Greece

Section 1

Maintenance Models and Policies

Chapter 1

Maintenance and Asset Life Cycle for Reliability Systems

Carmen Elena Patiño-Rodriguez and Fernando Jesus Guevara Carazas

Abstract

This chapter presents tools, methods, and indicators, in order to develop a successful and modern maintenance program. These are based on reliability engineering that improves the reliability of a system or complex equipment. Frequently, the industry implements maintenance schemes, which are based on equipment's manufacturer's recommendations and may not apply changes throughout the asset life cycle. In this sense, several philosophies, methodologies, and standards seek to assist this process, but most of them do not take into consideration their operation characteristics, production necessity, and other factors that are regarded as being important to one's company. This method is based on the analysis of preventive component replacements and the subsequent critical consequences. These analyses may be used as a decision-making tool for defining component replacement decisions. In this chapter, the first section introduces and justifies the importance of this topic being approached from the perspective of asset management. Next, it discusses key maintenance concepts and techniques, with the aim of establishing the foundation of a maintenance management. The purpose of the final section is to present a maintenance strategy model, and it presents the findings of the case study about model implementation at home cleaning service company.

Keywords: maintenance asset management, reliability management, maintenance optimization procedure

1. Introduction

Industry implements maintenance schemes based on equipment manufacturers' recommendations that might not be able to generate positive changes throughout the asset's life cycle. For instance, some diesel engines are designed to operate in Europe. Maintenance tasks need to be adjusted to operate in the South American tropics. This also happens with automatic transmissions. These tasks, sometimes, are neither adjusted nor improved. In this sense, several philosophies, methodologies, and standards seek to assist this process; however, most of them do not take into consideration their operation characteristics, production necessity, and other factors that are important to a company.

This chapter presents a modern maintenance strategy proposal aimed to comply with the ISO 55000 series of standards. These strategies are needed to develop a successful and modern maintenance program. In doing so, an appropriate maintenance strategy ought to be defined that will form the foundation for ensuring a high reliability degree of operating production systems. The challenge is to restructure maintenance strategies and, hence, to guarantee a high reliability level of the production system operations. The strategy presented herein was validated in a transport truck public company's policy regarding operational excellence and asset management, achieving satisfactory results.

The concept of "maintenance" in the industry has evolved in the last two decades. It is no longer seen as an expense or a team simply responsible for replacing production system components. Now, maintenance is considered an indispensable activity which guarantees not only the availability and functionality of a system or a component but also the high quality of the goods and services produced [1]. Likewise, in the early years, maintenance has solely been the responsibility of mechanical and electrical engineers. However, managing maintenance activities has become a multidisciplinary and far-reaching task within the organization. Maintenance directly impacts levels of production, budgets, timelines, and forecasted profits. Maintenance also increases the lifetime of equipment and ensures acceptable levels of reliability during usage. This occurs in every step from preventive maintenance through redesign. Moreover, operation teams must adapt to the specifications of each piece of equipment and each industrial need. Critical equipment may not have been manufactured uniquely for the organization's specific facilities, operators, or supplies. Additionally, proper management of equipment lowers operational costs. It reduces energy consumption, maintenance resources themselves (such as spare parts and labor), and risks to system operators, facilities, and production. Overall, managing maintenance activities results in savings for the organization.

However, production and engineering leaders focus on generating, modifying, and restructuring maintenance plans. Organizations consider the following questions: "Where to begin?" "Do we need to restructure the department of maintenance?" "Is it necessary to create management for this field?" "What kind of structure should we use?" and the like.

The objective of this chapter is focused on identifying three fundamental pillars for highly reliable systems: managing information, creating indicators, and restructuring preventative maintenance plans. These concepts aim to support production and maintenance managers in decision-making processes. They equally intend to support individuals and organizations seeking excellence in maintenance management practices in terms of facilitating decisions based on information with principles of excellence.

This chapter is organized as follows:

Section 2 provides a brief history of maintenance management and the definitions, related terms, and fundamental concepts.

Section 3 presents the proposed maintenance strategy model and the main results and analysis stemming from a study case. Lastly, conclusions are drawn in Section 4.

2. Literature review

Several scholars postulate the necessity of creating an integrated maintenance management system. This management system should aid the decision-making process and include some level of forecasting acknowledging the inevitability of occasional failure [2]. In other words, effective management requires systems and tools to predict the reliability of production systems. Predicting failures or defects with a high degree of certainty allows the operator to manage logistics and resources necessary to make interventions with the least impact on production [3, 4]. Moreover, it is necessary to clearly identify the goals of maintenance management within the organization, which must fully align with those of corporate management. Thus, maintenance decisions ought to be strategically framed within the corporate mission [5, 6].

The major changes to maintenance strategy are due to a need for more efficient production lines. The latter was sparsely automated, of low complexity, and only corrective in nature before the Second World War. Performed literature review reveals that this era of maintenance strategy came to a close in the 1950s [7]. From this point until the 1970s, the so-called second maintenance generation was developed. This era was characterized by the implementation of process planning, the advancement of technology, and more complex equipment. It also marked the beginning of industrial automation. In short, maintenance was based on well-defined cycles of spares, replacement, and reconstruction of equipment. In the pursuit of high reliability levels, these cycles became very short and ultimately drove an increase in maintenance costs [8].

The third generation of maintenance was marked by the influence of the aeronautical industry and their particular maintenance needs as required by the Federal Aviation Administration, in particular with the start-up of the Boeing 747 aircraft [9].

This change to maintenance brought financial hardships, which is why United Airlines formed a team to evaluate potential means of developing a new preventive maintenance strategy so as to find the balance between safety and costs in the operation of commercial aircrafts [10]. These changes have been considered and implemented in maintenance planning and activities for large aircrafts up to now. The circular advisory, maintenance steering group (MSG-3), presented a methodology for developing scheduled maintenance tasks and intervals acceptable to the regulatory authorities, operators, and manufacturers [11]. Years later, the MSG-3 gave rise to the current methodology of reliability-centered maintenance (RCM) [12]. The same was characterized by increasing demands in terms of quality for products and services alike. This in turn gave rise to standards and regulations that called for implementing changes in the traditional way of operating production systems. In the never-ending quest to establish optimal conditions for preventive maintenance, the probability and reliability studies of the aeronautical industry were applied in the production industry, as well. These early reliability studies were initially applied to providers of electrical energy in thermonuclear power plants, soon to be followed by the gas and petroleum industry, and was finally adopted and implemented by the general industry [13, 14].

The application of maintenance-specific reliability concepts characterized the fourth generation of maintenance standards, which in turn exemplified high-quality production and described the need for addressing operators' safety, as well as the proper operation of the equipment and the protection of the environment [15]. The fourth generation wanted to keep sight of resource optimization or the production of high-value goods. Value is defined as performance over cost and is presented in Eq. (1):

$$Value = \frac{Performance}{Cost} \tag{1}$$

Currently, the concepts of risk assessment and operational excellence were incorporated as a target of maintenance activities to minimize system failures and to guarantee reliability and availability. This maintenance stage was characterized by the implementation of risk-based maintenance techniques, such as risk-based maintenance (RBM) and risk-based inspection (RBI), which take the risk of an issue into account for the entire maintenance processes. At the same time, it was influenced by the new management standards, namely, asset management and facility management [16, 17].

The Federation of European Risk Management Associations (FERMA) states that it would be practically impossible to encompass every technique for risk analysis in a single standard and, likewise, impossible to resolve all problems with only one method. For this reason, each industry must adapt or develop its own method instead of trying to find a single general method. In other words, the methods implemented must consider the actual operation and asset failure, as well as the operating environments thus far, since all these aspects affect its performance.

2.1 Concepts and definitions

The British Standard Glossary defined maintenance as "the combination of all technical and administrative actions, including supervision actions, intended to retain an item in, or restore it to, a state in which it can perform a required function" [18]. In addition, maintenance is a set of organized activities that are carried out in order to keep an item in its best operational condition with minimum cost required. Likewise, maintenance tasks are defined as "Sequence of elementary maintenance activities carried out for a given purpose. Examples include diagnosis, localization, function check-out, or combinations" [19].

Preventive maintenance is the performance of inspection and/or servicing tasks that have been pre-planned or scheduled for specific points in time in order to retain the functional capabilities of operating equipment or systems [20, 21]. Other standards such as ISO 13372:2012 [22] define preventive maintenance as "maintenance performed according to a fixed schedule, or according to a prescribed criterion, that detects or prevents degradation of a functional structure, system or component, in order to sustain or extend its useful life."

Corrective or reactive maintenance is carried out after fault recognition and intended to put an item into a state in which it can perform a required function [23]. This maintenance policy is also called failure-based maintenance because the asset is operated until it fails.

Predictive maintenance refers to the routine inspection of equipment, machines, or materials to prevent a failure. It is a type of proactive maintenance that focuses on determining the potential root causes of machine or material failure and dealing with those issues before problems occur. It is achieved by the measurement of some physical or performance variable [24].

Robert Davis defined asset management as "a mindset which sees physical assets not as inanimate and unchanging lumps of metal/plastic/concrete, but as objects and systems which respond to their environment, change and normally deteriorate with use, and progressively grow old, then fail, stop working, and eventually die" [25].

Table 1 shows additional important concepts of maintenance management for reliability systems, in which the following four factors are recognized:

a. Equipment has a life cycle

b. Maintenance management is as important for those working in finance as it is for engineers

c. It is an approach that looks to get the best out of the equipment for the benefit of the organization and/or its stakeholders

d. It is about understanding and managing the risk associated with owning assets such as equipment

2.2 Fundamental aspects in the maintenance strategies

Maintenance decisions are of diverse natures and, depending on the level of impact, require proper identification and ranking. This is the starting point to develop suitable management policies and bring assertive strategies of reliability. Decisions associated with production maintenance are of four levels:

a. Instrumental (dispatch)

b. Operative

c. Tactical

d. Strategic

The strategic level incorporates the top direction of the organization and the maintenance implementation with tangible results in a time frame upward of 2 years. These decisions require important investments of resources and market studies, opportunities, and returns on investment.

In operations and production, tactical decisions generate results within several months to 1 or 2 years. Tactical decisions are made by management and mid-level management, involve project modifications, and often are associated with important investments. Operation-specific decisions have immediate impact (from several days to a few months) and are made by technical personnel that do not require changes and investments in the operational budgets. Instrumental or dispatch decisions are also made by technical personnel. The costs related to these decisions are considered in the plans pertaining to preventive maintenance, and their impact is reflected in hours. The maintenance activities related to these decisions are called adjustments.

Often, the governing bodies of the industries only stop to consider the need to restructure their departments or maintenance processes when faced with frequent expensive failures or costly downtimes that cause significant production losses. In addition to the above, performed research indicates that implied processes of documentation and registration processes are precarious, even though, in many cases, significant sums of money have been invested in information systems. When

Term	Definitions
Availability	Ability to be in a state to perform as and when required, under given conditions, assuming that the necessary external resources are provided
CBM	Condition-based maintenance: preventive maintenance which includes a combination of condition monitoring and/or inspection and/or testing, analysis, and the ensuing maintenance actions
CMMS	Computerized maintenance management system: a system that can provide important information that will assist the maintenance management in planning, organizing, and controlling maintenance actions
CMMS	Computerized maintenance management system: a system that can provide important information that will assist the maintenance management in planning, organizing, and controlling maintenance actions

Table 1.
Concepts for maintenance management for reliability systems [26].

confronted with these loss-potential scenario initiatives to strengthen and structure, the corresponding maintenance departments are taken.

The following are the first steps to properly establish the maintenance requirements inside the organization to guarantee high reliability, equipment availability, and compliance with operational and environmental risk regulations.

It has to be noted however that discussing the performance evaluation of a production system without having prior implemented a maintenance information system may lead to inherent failures. Indeed, an MIS is a tool in which failures, time interventions, spare parts, etc. are saved, treated, and processed in order to inform maintenance managers and facilitate decisions. Although there could be other tools to evaluate the performance of the production equipment, the maintenance information system is where the key indices are considered and integrated with the general maintenance strategies.

Overall, a maintenance information system has four main functions:

a. Collect data

b. Support engineering decisions

c. Record interventions

d. Plan for spare parts and equipment expenses [27, 28].

The MIS can be integrated with a general computerized maintenance management system (CMMS).

The following sections introduce on the one hand its basic aspects and on the other hand highlight the means of using one for performance evaluations and, subsequently, decision-making processes.

In order to have organized and conscientious data collection, it is imperative to define the following:

a. The critical assets

b. The failure

c. The desired capabilities and limits according to the functions for which the assets were designed

The reason therein may be associated but not limited to the fact that the plant could have hundreds of assets which could result in useless, inefficient work. Data collection should begin with an organized set of information which must include static information, such as:

- Hierarchy classification
- Nameplate information
- Processes and instrumentation diagrams (PIDs)
- Assembly and spare part drawings
- Functional analysis charts

- Catalogs, technical bulletins, etc.

Once static data is collected, it is important to record information related to failures and interventions. It is at this point where the record of work orders and failure report may be used as it may be regarded as the foundation of the availability and maintenance indicators incorporating essential information concerning financial planning and evaluation. The correct filing of work orders should include at least the nametag of the asset, time records, workforce, downtimes, spare parts, and detailed descriptions of activities and operative windows.

Similarly, a fault report should accurately describe the type, nature, and time the fault was observed and, if it is already cataloged, put the fail mode number or tag. There are international standards such as the ISO 14224 describing general guidelines to report faults and tag them. One of the biggest benefits of recording the failures according to the international standard is the ability to share and use information to estimate failure rates. An example of the failure rate prediction and database is in the OREDA Handbook for the offshore oil and gas industry [29, 30].

A work order is the main tool that allows recording fault information. It begins with a planning process in which the workforce, deadlines, procedures, and route sheets are established. The work order continues with a programming stage where the precise dates and the maintainer's ID are selected, and, after this, the work order is executed and closed in the information system. This last step may be regarded as the triggering point and the interface to the real world as it gives rise to all these processes, since it documents and depicts in the equivalent data record, the KPI's computing, evaluation, and the provision for making decision [31–33].

Even so, work orders and failure reports are not enough. If only work order data is tracked, it is difficult to establish tendencies, averages, and alerts. As such, it may not be possible to establish equipment-specific degradation levels as well [34]. This is when quantitative variables become necessary because they indicate the performance state of the equipment. These quantitative variables come from sensory devices such as gauges, thermometers, pressure and temperature transducers, flow meters, gas detectors, vibration sensors, etc. It is important to highlight the fact that a quantitative variable could be useful only if the correct functions of the equipment and their parameters are well established. This may be demonstrated by using the PF curve [35]. In some plants, SCADA and DCS are commonly found where the variables can be analyzed remotely and stored, and, in many cases, they are only used for operational purposes. In brief, quantitative and qualitative maintenance and cost data are necessary to evaluate the performance of any asset or piece of equipment.

3. Result and analysis

3.1 Proposed maintenance strategy model

The model proposed strategically incorporates the "better practices of maintenance management" in order to achieve operational excellence in the framework of the international standard ISO 55000:2014. Better practices in maintenance management have the following attributes: they are realistic, specific, achievable, and tested in the industry; they contribute in making maintenance more efficient and profitable, while optimizing operation costs and improving equipment's reliability. Authors have equally postulated an overall improved level of satisfaction and motivation among personnel [36].

In order to identify the most relevant indicators facing a company's maintenance strategy, it is necessary to distinguish between effectiveness and efficiency. For maintenance purposes, effectiveness measures the health of equipment, while efficiency measures the state of the equipment in comparison with the effort and resources needed to maintain that state (**Figure 1**).

Once relevant assets, work order flows, and related indices for efficiency and effectiveness are identified, it is possible to discuss maintenance optimization and economic evaluation by considering how to predict failure rates with quantitative and qualitative data. Although a common maintenance information system does not include tools such as FMEA, RCM, and data analytic packages, it is important because it allows users to analyze information and the subsequent decision-making. In conjunction with the data collected, different maintenance strategies (preventive, periodic, and predictive) can be analyzed and compared.

The model incorporates the building blocks of the ISO 55000, PAS 55, and ISO 39001 series of standards and promotes the development in three stages, namely:

a. Planning

b. Process design

c. Maintenance management

It is worth noting at this point that all stages mentioned above integrate personnel, processes, and the equipment in an improvement cycle, as described later.

The proposed model is based on the requirements of an asset management system set out in the international ISO 55000 and ISO 55001 series of standards while considering aspects of the ISO 39001 standard. The latter addresses the fundamentals for developing a road safety management system, such as shown in **Figure 2**.

3.1.1 Phases of the maintenance strategy model proposed

The development of a maintenance strategy model could be a long process, and it could depend on each productive system or company guidelines or even real context. Due to the latter, it could be difficult to run into a particular methodology which comes over all necessities. It has the intention to bring up some guidance for the achievement of a maintenance strategy. Although this guidance has a general focus, it was applied in a truck fleet and consequently may be more specific in some areas/ideas.

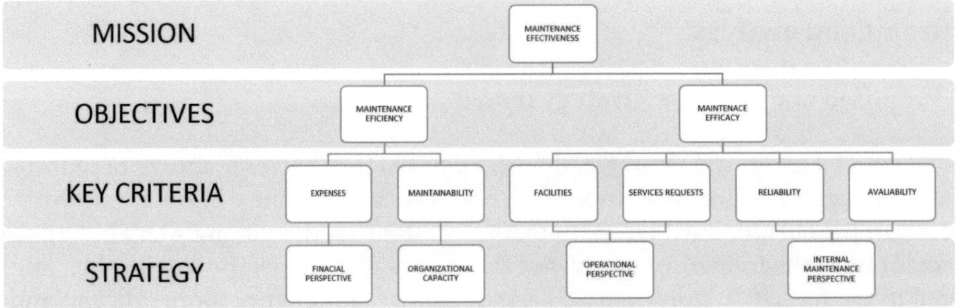

Figure 1.
Maintenance effectiveness indicators.

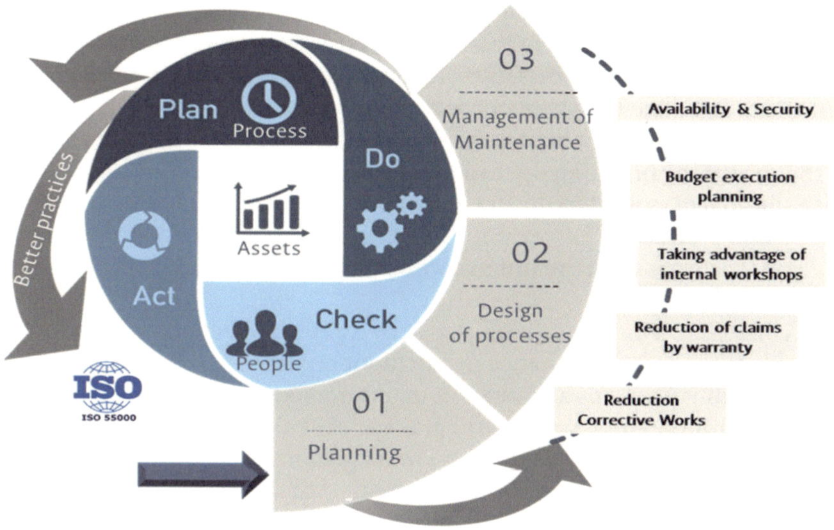

Figure 2.
Excellence model.

For the reason that the proposed model is based on the guidelines of international standards, it is suggested that the aforementioned fundamentals be considered during the execution of each stage of the model. These stages are developed through frequent interaction of the staff (tactical and operational level) with the maintenance processes and finally the interaction with top-level management in charge with setting up the company's strategy.

In order to develop the model, the following stages are considered:

a. Planning

b. Process management

c. Data collection

d. Process evaluation

e. Baseline development

f. Feedback

The development of these phases results in a maintenance management process that is lately generating value for the organization:

- Planning maintenance management: In this stage, the current state of corporate maintenance management is analyzed, considering the mission of the organization, the identification of the vision and mission, and the core focus of the business. The strategic indicators have to be considered and raised and defined in the "performance evaluation" with the aim to establish the starting point and address the processes to higher levels of excellence. In this stage, activities this stage, activities such as budget planning and execution, maintenance plan checking, resource planning and spare part planning, predictive task managing, and inspections, among others, ought to be considered [37].

- Process management: This stage introduces the current activities developed in maintenance management teams. Contextualization is necessary when the organization is in the process of restructuring. This applies in cases such as a new maintenance management team or if an asset management process is being structured. Maintenance management must be studied and reconsidered in the context of pursuing operational excellence. This full process is developed with the goal of not interrupting daily operations by structuring of the new maintenance strategy. Developing this stage often initiates documentation that becomes the basis of the maintenance strategy and endures over time.

- Data collection: The objective of this activity is to collect all information available from the maintenance department regarding assets such as technical sheets, roadmaps, plans and current maintenance frequencies, manuals of parts and components, spare part catalogs, checklists and inspection formats, inventory of components and assets, management procedures and technical processes, considerations, results of current indicators, and requirements.

This stage can be complex depending on the organization. It is the authors' view that even if one does not manage to complete the whole survey, one should continue onto the other stages. It is deemed that this point should not become a "dead end" or a bottleneck of the process. In the future, one may update it using on the one hand the information system and on the other hand, information from providers, among other sources. In order to accomplish this stage, it is necessary to devote workforce and work plans so as to do all data collection tasks.

- Information assembling and analysis: This stage, as its name implies, consists of the organization of information. A large part of its success relies on focusing on the amount of information collected in the previous stage. At this point the information is organized in order to eliminate irrelevant matters that are of no value to a company's strategic objectives. As such, the needs of storage, capture, and updating are defined. Additionally, it is the authors' view that corporates information systems such as ERP, EAM, CMMS, or simply databases ought to be further developed and incorporated for measuring purposes.

Once all the relevant and necessary data have been compiled, the evaluation of the characteristics of the data that incorporates, among others, the identification of the information needs, the update or the creation of formats and the feedback of the new processes, if necessary. It is the authors' view that this stage should mark the participation of the information technology (IT) teams, which will define the most appropriate computational tools to load the information systems. The vast majority of robust information systems communicate in a friendly way with database files, since they are well structured by the IT team. With regard to maintenance activities, this point should equally identify any shortages such as maintenance plans, frequency adjustments, elimination of assets not in use or already written off, equipment not incorporated, etc.

- Process evaluation: At this stage, the existing processes in the maintenance management system are surveyed. It is worth noting that from a study [38] of about 14 companies in the mineral extraction sector, only 4 companies had documented processes associated with maintenance activities, the vast majority of which were related to purchase processes. Once processes have been documented, the effectiveness of these ought to be analyzed, by identifying the inputs and outputs of each and their means of interrelation. Emphasis ought to be placed on the structure needed to capture the data that can generate management indicators and establish controls. At this stage, the team must work more closely with the quality team in order to verify the documentation processes of the international standard [39].

This phase will bring to light any potential needs pertaining to the modification and the generation of new processes that are accompanied by all the documentation and methodology. This is detailed in previous sections:

- Baseline development: It is to generate a minimum state in the equipment that thoroughly satisfied its primary functions. This state must have an acceptable level of reliability for them to operate safely. This point is possibly the most complex because it consolidates the previous stages in order to generate a baseline of work. This is where the mission and objectives of maintenance management are well-defined. The KPI classes are defined to follow up on new and consolidated ones. Only with the fulfillment of this stage do operation structures change for improved productivity. Also, savings begin as a result of the elimination of redundant or unnecessary processes. The generation of the baseline gives a solid start to the knowledge of maintenance needs in the organization [40]. To complement this stage and its results, it is necessary to communicate with the personnel involved and responsible for production, by identifying improvement opportunities, defining the actions to be implemented, and clearly establishing the requirements necessary for an adequate implementation of the strategies.

- Feedback: As a fundamental part of the operational excellence model, the team will be confronted with the strategic objectives of this project. These activities are monitored, and improvement plans are established (preventive or corrective) in accordance with the traditional process of continuous improvement presented by the Deming Prize Committee in 1950 [41].

3.1.2 Maintenance process design

This section presents the necessary processes pertaining to maintenance management which are presented under the guidelines of the international standards [42], such as:

- EN 16646 Maintenance—Maintenance within physical asset management

- ISO 55001 Asset management—Management systems—Requirements

- ISO 9001 Quality management systems—Requirements

According to the ISO 9001 standard, a process is the set of mutually related activities that interact, transforming input elements into ISO 9001 output elements. In maintenance management, the input elements are usually associated with operational demands, requests for intervention over assets, results of internal and/or external audits, needs for the maintenance of assets, and customer requirements, among others. In order for these to be transformed into maintenance plans, preventive, corrective, or improvement actions are aimed at achieving strategic goals. The objective of designing a process for maintenance management is to achieve compliance with the specifications required by all interested parties (customers, shareholders, related entities) such as costs, quality, flexibility, availability, reliability, maintainability, operation times, environmental regulations, safety, and health, among others.

Consequently, it involves making strategic decisions regarding human resources, machinery, tools, materials, infrastructure, methods, and technologies to be used. In general, it is the authors' view that it is necessary to design or redesign a process in the following cases that involve:

- Important modifications in the requirements
- Quality problems
- Priorities of the organization have changed
- Altered demand
- Performance indicators not reaching the expected results
- New processes or technologies used by competitors
- Important changes in the inputs or in cases where their availability has changed significantly

The issues mentioned above are derived from a full analysis of the internal and external context of the organization, a necessary requirement to implement standards ISO 55001 and ISO 9001 [43].

Designing a process involves the definition and systematic management of all processes and their interactions, for which analysts can use visualization tools such as process maps, information flowcharts, and task lists by activity. These tools help under a process management approach to establish the following:

- The existing processes
- The relationship between processes
- Strengths and weaknesses
- Easier operations
- Activity and operation integration
- Activities and tasks which might be eliminated or do not add value
- Delivery delays or issues
- Communication flow issues

Bravo C. in his work [44] expresses that "... Process management is a discipline that helps the management of the company to identify, represent, design, formalize, control, improve and make more productive the processes of the organization to win customer confidence. The organization's strategy provides the necessary definitions in a context of wide participation, where process specialists are the facilitators..."

At the same time, the cited work presents a four-cycle framework for the integral change management. These cycle stages are listed in order as follows:

1. Strategy design

2. Visual modeling

3. Process intervention

4. Useful life management

The four cycles mentioned above incorporate new practices and require a high commitment from all bodies. Based on the strategy and on a preliminary analysis of maintenance processes, it is possible to build a process map, which must be circulated to all organization personnel. A process map provides a global–local perspective, grouping each process into strategic, key, or support. The design of a process map depends on the context in which each organization is developed under the following criteria:

- Strategic processes: They are identified at the top of a process map, and their objective is to plan the strategies of the organization, make the relevant plans, and provide feedback to the other processes. In maintenance management these processes are related to the planning of the activities to be carried out, in accordance with the work orders that are generated, the monitoring of performance indicators, and the generation of policies to improve the results. The BS EN 16646 standard recommends considering the following processes: planning of maintenance activities, management and development of resources, creation of maintenance plans, monitoring and continuous improvement, evaluation and control of risk, and decisions regarding the portfolio of assets.

- Key or business processes: They are pinpointed at the center of a process map; they are derived directly from the organization's mission. In a maintenance department, the processes involved here correspond to the execution of preventive, corrective, or predictive plans from the implementation of asset management to the generation and scheduling of work orders and the supervision of actions in the operation plant. According to the maturity of the organization, this layer may also include the processes of acquisition of physical assets (if they exist in the market) or manufacturing physical assets (if they do not exist in the market in acceptable economic conditions). This may also include updating or improving assets for higher value throughout the global life cycle of the assets, taking out of service, and/or withdrawal of assets when their utility is worn out.

- Support processes: They are identified at the bottom of a process map and support the entire organization in aspects that are not directly related to the business, but it is necessary to convert the strategies into concrete activities. In maintenance management, this includes the communication protocols, inspection and diagnosis of the assets, and the monitoring of processes designed to achieve the organizational objectives. One may consider processes for resource management (human, information, materials, and tools) and information management (CMMS).

The relationship of the processes depends on the context of the organization, as well as the specifications of the associated procedures and the detail that the instructions and records must possess (see **Figure 3**). Once the processes and the way they are related are identified, the specific procedures of the key activities must be characterized and defined. This equally includes the instructions for technical operations (inspection routines, road maps, among others) and the formats of the records necessary for the analysis of data (asset resumes, fleet profile, failures, failure modes, frequent causes), as will be explained later.

The characterization involves documenting each of the processes designed for management, identifying the inputs, outputs, and activities in each of the stages of the improvement cycle proposed by Deming in 1950 (PDCA). For their part, the procedures detail the sequential steps to properly develop the processes, which in some cases are stored as part of a process manual. Roadmaps detail the procedures considered for maintenance management and are used at technical levels. These tools must show records of their execution necessary for the monitoring of activities and the collection of performance indicators (KPI).

3.1.3 Maintenance operative process management

The management of operational processes from the tactical perspective refers to the need for a system that allows the administration of work, materials, and resources, in order to gain control over the maintenance processes while requiring planning and programming that include an established order of work, equipment stops, and the creation and development of preventive and predictive maintenance plans. Within the framework of this management, the performance of the work team must be measured at each level, and the performance of aspects such as the implementation of lubrication routines, inspection, condition monitoring, and activities for the prevention of failures must be evaluated. The scope of process management incorporates five areas with defined global scope.

Work management guarantees well-established planning and programming that all tasks are planned at least 24 hours in advance and programmed with a week minimum margin, except emergency work. The adequate administration implies the existence of criteria for the creation and programming of work orders, which are used and respected, wherein the work flow is continuous and is not hindered by material or resource problems and in case of delays there are no major disturbances of the schedule. The latter implies that these are contained in 2 to 4 weeks of work. The indicator of worker efficiency is high, which leads to high staff performance.

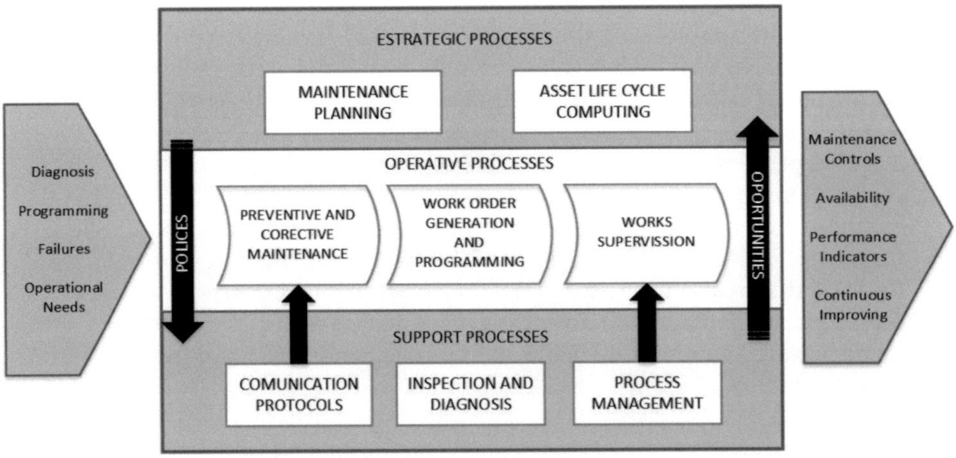

Figure 3.
Process map of maintenance asset proposed.

For the adequate workflow, the design of the work order is necessary, which must act as a transversal mechanism to guarantee compliance with the Deming Cycle (PDCA) in the flow of maintenance activities. The work order must be standardized as a document that calls on the completion of a task or set of tasks and serves, among others. It should be considered as the nucleus for the compilation of data, for the attention as a whole or for the attention of individual components and their processes. The work order becomes a starting point for the control mechanism since it transmits information about the work carried out, the start dates, estimated completion, and actual completion.

The work order flow must involve all the maintenance and operation personnel. It shall reflect the prioritization of the needs where the most critical and urgent must be dealt with first. Another suggested point is to establish a hierarchical limit in the execution of the work order. Therefore, in this stage it is necessary to define among four options: a) include actions at the system level, b) include actions at the subsystem level, c) include specific part tasks, or d) include inspection routines. This hierarchical limit will allow the tracking of work orders within the operational model of excellence. In addition, a work order must be allocated to the personnel in charge, detailing, among others, the materials, resources, previous analysis of the situation, and static data such as manuals, inspection routines, catalogs, etc. Furthermore, it must give space for the order of opening, planning, programming, and finalization. In general, the cycle time of a maintenance work order can be reduced by incorporating the following activities:

- Management of main stops: The maintenance management involves scheduled stops up to 6 months in advance and a precise definition of the scope of the work to be executed, giving enough time for realistic fulfillment of the objectives. This implies managing the process, formalizing scheduled stops, high involvement by production, engineering, maintenance, and processes. In the period of fulfillment of the scheduled major stop, attention is only given to emergencies.

- Management of materials and resources: The availability of material and resources is solved with automated inventory controls that are part of the maintenance management information system and by stock levels supported by the economic analysis of internal maintenance. Resource management is based on the history of materials and resources, generation lists, vendor databases, inventory monitoring, and inputs.

- Management indicators: Evaluating performance is part of the day-to-day process. Key indicators characterize costs in terms of quantity, type, area of origin, materials and resources, and work order. The management indicators should contemplate, measure, and obtain information on the company, plant, departments, improvement team, and work teams. The process indicators are directed to be effective, and external and internal benchmarking is used to lead the process.

- Reliability management: For operational processes to achieve a high reliability degree, it is essential to use CMMS/EAM as a tool for making optimized decisions, along with the experience of the staff. The use of the systems includes a diverse area of disciplines such as engineers, planners, and different work teams. The analysis of the condition is linked to the monitoring and preventive maintenance activities completed in all areas. The frequencies and activities of the maintenance routines are refined through the feedback of the work order and a root cause analysis of the failures.

- Planning and programming: It is the authors' view that adequate planning and programming should include short-term activities in the planning and scheduling of preventive maintenance. Activities of greater complexity can be addressed through root cause analysis. Likewise if, say, 80% of the total activities is scheduled in adequate time, then this may be regarded as demonstrating a stable maintenance operation. Another important point is to try planning in the long term and scheduling in the short term as much as possible.

Requirements for proper planning and programming include understanding the need to respond, properly preparing a work order with appropriate prioritization, and integrating operations to reduce programming delays due to nonavailability of the asset for maintenance [45]. Planning and programming involve:

- Assigning a programmer and planner to review the pending work and coordinating modifications in the allotted time
- Establishing roles, responsibilities, rules, and lines of authority between planners and programmers and maintenance and operations
- To assign engineering as a support to planners
- Conducting daily meetings between programmers and operations to align the needs of both parties
- Measuring delay times and operating hours
- Holding meetings to level the needs of programmers and planners alike
- Establishing a defined level of service for materials and resources
- Notifying maintenance or purchasing leaders of material and resource needs
- Considering routes for maintenance personnel and analyzing the tasks and frequencies of the assigned activities

The improvement cycle of planning and programming begins with an analysis of the existing maintenance plans and ends with a new plan, whose effectiveness is measured from the mean time between failures classified as critical systems. The implementation is progressive from the identification of the most critical systems, considering relevant indicators such as mean time between failures (MTBF) as input variable. The improvement process at its starting point cannot ignore the recommendations of the manufacturers.

3.1.4 Critical maintenance task (CMT) list and regular maintenance task list

The process to generate critical roadmaps and regular route sheets for maintenance tasks begins in accordance with asset ranking, that is, the severity of the impact of their failures within the production process. The hierarchical level of assets is defined in the generation stage of the baseline.

Roadmaps, by definition, are documents designed to direct maintenance activities by minimizing the level of human error on the part of operators. They were developed by the aeronautical industry in the 1970s within the framework of technical recommendations pertaining to reliability in maintenance [46]. Roadmaps direct maintenance activities from preventive to predictive and even corrective, with the aim of reducing human error during their development and, thus, guaranteeing high reliability in complex and high-risk systems.

Implementation of roadmaps allows the development of technical benefits in integral maintenance management. They detail the activities, procedures, tools, and spare parts necessary for the execution of each of the activities scheduled in stated preventive maintenance plans and are based on the technical specifications recommended by the manufacturer [47]. Based on the work of the aeronautical industry, the reduction of error can be concentrated as shown in the **Table 2**.

The maintenance tasks can be classified in the main groups: corrective, preventive, and predictive (condition-based and condition monitoring) [48]. Souza and Guevara present two tables that can help determine the main causes of mechanical failures based on RCM studies [49].

3.2 Case study

The proposed model was implemented at an organization with 54 years of experience in providing home cleaning services and complementary activities in the city of Medellin and five nearby municipalities. The company has 767,668 users, among the residential, commercial, and industrial sectors. Service delivery in the residential sector is carried out twice per week (Monday–Thursday, Tuesday–Friday, Wednesday–Saturday), for a total of 104 services per year. Frequencies in the commercial and industrial sector may vary between 1 to 7 times per week, depending on the waste generation of each subscriber, which leads to a total of 104 to 365 collections per year. The main activities of the organization are collection and transport of solid wastes, sweeping and cleaning of roads and public areas, grass cutting and pruning of trees in public areas, and washing off roads and public areas. The range of services extends to the collection of special wastes, among which are waste generated at events and mass shows, points of sale in public areas, dead animals, construction and demolition wastes (C&D), hospital wastes, mattresses, vegetables, furniture, carpentry wastes, and collection (dismantling and installation) of public baskets.

As part of the solid waste collection strategy, the organization has a diversity of vehicles with different dimensions in order to access areas with adverse geographic conditions. To allow great maneuverability in limited-access roads, the company has model 2009 Kenworth vehicles with only two axes (simple) and smaller vehicles such as NPR model 1998 and 2012. In general, to meet the demand, the organization

Error location in flowchart	Definition
Scheduling (E1)	Wrong execution of either of the two tasks: identify next inspection or move to a location
Inspection (E2)	Not seeing a defect when one exists
Inspection (E3)	If human induced, due to either forgetting to cover area, covering area inadequately, or a scheduling error
Engineering judgment (E4)	An error in deciding whether the area in which a defect is found is significant or not
Maintenance card system (E5)	Arises because the work cards themselves may not be used to note defects on the hangar door immediately as they are found
Noting defect (E6)	The error is noted incorrectly or not noted at all

Table 2.
Potential errors in the inspection process.

has its own fleet broken down as follows for each type of service. Collecting wastes from hospitals are carried out by using three vehicles equipped with containers and UV light, to ensure crew condition and reduce biological risk. The vehicle fleet has two skid-steer loaders transported by dump trucks for the collection of C&D waste. The organization operates light vehicles (NPR) for transportation of baskets in poor condition for waste disposal. Besides, the vehicle fleet has two series of equipment that allow the provision of collection services for special containers. Both use a lifting system equipped in the back part of the truck which is called lifter. One of the series of vehicles is employed for collection of big containers, whereas another series is used for the collection of buried containers. An availability indicator is generated, based on the data of the information system reports and the work orders of the equipment maintenance activities. All of this for a 30-day operation period during 2018 is shown in **Figure 4**.

In **Figure 4a**, it is possible to observe the increase in availability of the fleet of light vehicles per quarters during 2018, from 78% in the first quarter to 89% in the fourth quarter showing signs of stabilization. This improvement may be attributed to the reduction of the occurrence of failures, which in turn is the result of the implementation of a preventive maintenance program, and critical maintenance tasks. These availability standards are appropriate for a program of operational excellence. During 2018 one may see that the month having the best behavior in terms of availability was November showcasing a growing trend due to the implementation of the excellence model.

Despite the improvement in terms of availability that was evidenced in this study, it is the authors' view to analyze the specific behavior of the fleet vehicles. This is because being an average, the availability of the light vehicles could be

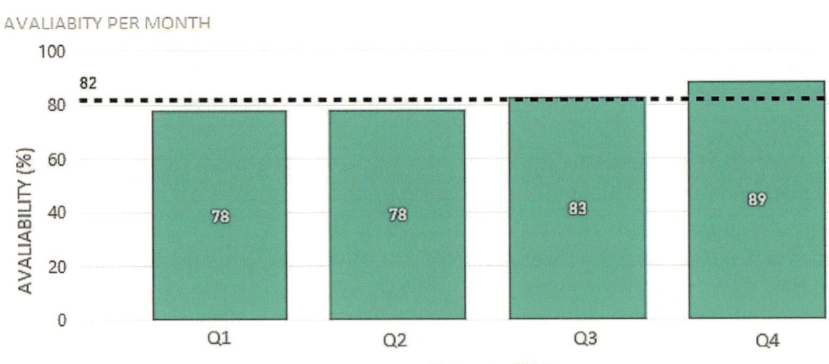

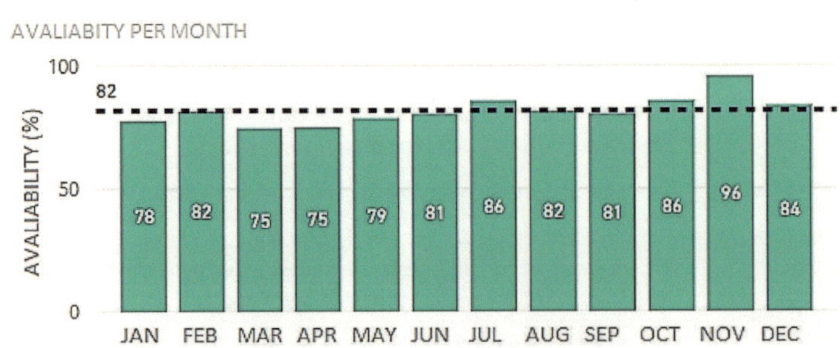

Figure 4.
Availability indicator for light vehicles per 2018 quarters.

affected by extreme values. **Figure 5** shows the behavior of each of the vehicles that make up the fleet. The vehicles that have worse average availability are #313 and #416. However, if the growth of the availability of these vehicles is observed, the positive impact of the implementation of the model of excellence can be verified.

To evaluate the impact of the excellence model implementation, availability indicators besides indicators like efficiency must be considered. Next, the data related to the maintenance cost is compared, through the execution of work orders. The amount of orders generated, their typology, and the effective cost paid for these works are taken into consideration.

It is pertinent to mention that each work order carried out carries with it the corresponding audit report, which gives information of the tasks executed in detail, the report of time units, and the quality of spare parts used. These data are essential for the administrative review phases, in case of any type of claim, and verify the agreement with the terms of the contract.

Additionally, the maintenance work of the first 10 months of the year 2018 is analyzed and compared with the same period for the year 2017. A reduction of 5.06% in the sum of the maintenance costs is evidenced. In addition to this, the average cost per work order generated was reduced by 12.94% thanks to the recommendations of good management practices in the processing and management of information (**Table 3**).

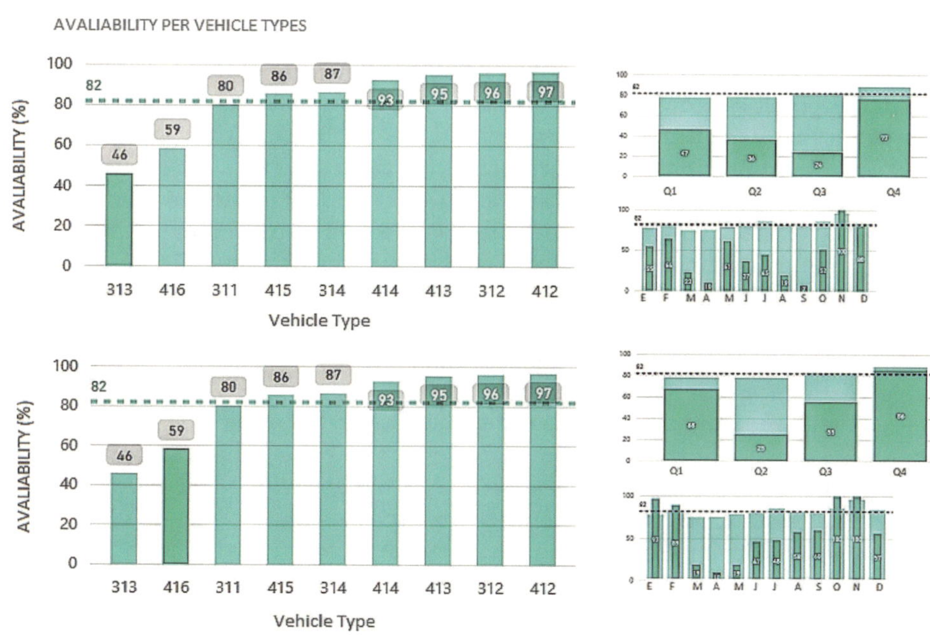

Figure 5.
Availability indicator for critical light vehicles.

	# OTS	% OTS prev	OTS value
2017	5755	6.23	−12.96%
2018	6277	10.12	
Difference (%)	−5.06%	3.89	

Table 3.
Budget execution in maintenance cost 2017 vs. budget execution in maintenance cost 2018.

In this analysis, an increase of 3.26% is observed. A fundamental part of this increase is due to the optimization of downtime, because with the operational excellence model, preventive critical tasks are programmed and executed in the same time periods of the corrective activities.

4. Conclusions

During the life cycle of an asset or a production system, different costs are incurred, which span the purchase (initial investment) to the operation and maintenance costs that guarantee productive and financially worthy outputs for investors. The life cycle cost corresponds to the costs of both investment and operations inherent to the useful life of the asset.

Development and implementation of management models, applied to the maintenance of equipment, often present results in periods that exceed 1 year. This model, based on the ISO 55000 standard, presented immediate results mainly by, firstly, defining maintenance as a strategic activity for the collective benefit of the organization and, secondly, collecting all necessary information pertaining to defining the critical maintenance needs, in such a way so as to guarantee the high availability of the assets.

In this chapter a maintenance strategy model of the asset life cycle is proposed. It has a direct influence on maintenance management regarding decision-making, as well as the planning of preventive tasks and analysis of the equipment's useful life. Positive results are obtained in the overall development of this maintenance model. It is possible to notice a reduction of costs in the global execution, and so the average cost per work order has been reduced too. At the same time, an increase in the execution proportion of preventive tasks has been achieved. These findings may help other to implement the model successfully, even though the tasks performed and the model itself remain in continuous analysis and improvement.

The common maintenance budget models only present a general sum of costs; it does not provide enough information for decision-making. These results confirm the association between cost control, technical decisions and physical interventions, which have been and exemplified, and therefore, in this document, a new way of disaggregating the cost has been suggested.

Frequently the summation is separated monthly to fit in with the scale of the time series. These estimators of central tendency allow the visualization of how data variability alters this tendency according to the behavior of the series. It is also evident that the median is less susceptible than the average.

The result of the implementation has shown an increase of the availability indicator and a reduction of the general maintenance costs.

These preliminary results that cover a 12-month period suggest that in the long term/medium term, the availability may reach the level demanded by the company and may guarantee stable operation with lower maintenance costs. Finally, it is important to highlight that, without the support of the general management of the organizations, the initiatives to achieve operational excellence, or an adequate management of assets, may fail, causing loss and discouragement.

Abbreviations

CMMS	computerized maintenance management system
CMT	critical maintenance tasks
DCS	distributed control system

EAM	enterprise asset management
ERP	enterprise resource planning
FAA	Federal Aviation Administration
FERMA	Federation of European Risk Management Associations
FMEA	failure modes and effect analysis
IT	information technology
KPI	key performance indicator
MIS	maintenance information system
MSG	maintenance steering group
MTBF	mean time between failures
OREDA	Offshore and Onshore Reliability Data
PID	processes and instrumentation diagrams
PDCA	plan, do, check, act
RCM	reliability-centered maintenance
RBI	risk-based inspection
RBM	risk-based maintenance
SCADA	Supervisory Control and Data Acquisition

Author details

Carmen Elena Patiño-Rodriguez[1*] and Fernando Jesus Guevara Carazas[2]

1 Department of Industrial Engineering, University of Antioquia, Medellín, Colombia

2 Department of Mechanical Engineering, Nacional University, Medellín, Colombia

*Address all correspondence to: elena.patino@udea.edu.co

References

[1] Pintelon L, Parodi-herz A, Pintelon L. Maintenance: An evolutionary perspective. In: Complex System Maintenance Handbook. 2008

[2] Castañeda DA. Toma de decisiones en la gerencia de mantenimiento: un enfoque desde la analítica aplicada. [Thesis]. Medelin: Universidad Nacional de Colombia; 2018. Available at: http://bdigital.unal.edu.co/64965/2/1036623898.2018.pdf

[3] Besnard F, Bertling L. An Approach for Condition-Based Maintenance Optimization Applied to Wind Turbine Blades. IEEE Transactions on Sustainable Energy. Jul. 2010;**1**(2):77-83

[4] Herrera IA, Nordskag AO, Myhre G, Halvorsen K. Aviation safety and maintenance under major organizational changes, investigating non-existing accidents. Accident; Analysis and Prevention. 2009;**41**(6):1155-1163

[5] Kelly A. Strategic Maintenance Planning. 1st ed. Oxford: Elsevier Butterworth-Heinemann; 2006. 284 p. ISBN: 10:0-75 066995-0

[6] Atrens A, Murthy DNP, Eccleston JA. Strategic maintenance management. Journal of Quality in Maintenance Engineering. Dec. 2002;**8**(4):287-305

[7] Moubray J. Reliability-centered Maintenance. 2nd ed. New York: Industrial Press Inc.; 1997. 440 p. ISBN 0-8311-3078-4

[8] Carazas Guevara FJ, Souza GFM. Risk-based decision making method for maintenance policy selection of thermal power plant equipment. Energy. Feb. 2010;**35**(2):964-975

[9] Nowlan FS, Heap HF. Reliability-centered Maintenance. San Francisco: United Air Lines Inc.; 1978. 515 p. [Report]. Available at: https://apps.dtic.mil/docs/citations/ADA066579 [Accessed: 07 October 2018]

[10] Smith AM, Hinchcliffe GR. RCM–Gateway to World Class Maintenance. 1st ed. Oxford: Elsevier Butterworth-Heinemann; 2004. 336 p. DOI: 10.1016/B978-0-7506-7461-4.X5000-X

[11] Cranfield University. Maintenance Steering Group-3 (MSG-3)–SKYbrary Aviation Safety. In: SKYbrary. [Online]. 2017. Available at: https://www.skybrary.aero/index.php/Maintenance_Steering_Group-3_(MSG-3). [Accessed: 06 January 2019]

[12] Guevara Carazas FJ, Marthade Souza GF. Reliability Analysis of Gas Turbine2012. pp. 189-220

[13] Cooke FL. Maintaining change: The maintenance function and the change process. New Technology, Work and Employment. 2003;**18**(1):35-49

[14] Rausand M. Reliability centered maintenance. Reliability Engineering and System Safety. May 1998;**60**(2):121-132

[15] Khan FI, Haddara MM. Risk-based maintenance (RBM): A quantitative approach for maintenance/inspection scheduling and planning. Journal of Loss Prevention in the Process Industries. 2003;**16**(6):561-573

[16] ISO 55000. Asset management – Overview. In: principles and terminology. 2014

[17] ISO 44001. Collaborative Business Relationship Management Systems–Requirements and Framework; 2017

[18] BS 3811. Glossary of Terms Used in Terotechnology; 1993

[19] BS EN 60300-3-11. Dependability management — Part 3-11: Application

guide — Reliability centred maintenance; 2009

[20] Federal Standard 1037C. Telecommunications: Glossary of Telecommunication Terms. 2000. [Online]. Available at: https://www.its.bldrdoc.gov/fs-1037/fs-1037c.htm. [Accessed: 07 January 2019]

[21] ISO 14224. Petroleum, Petrochemical and Natural Gas Industries — Collection and Exchange of Reliability and Maintenance Data for Equipment; 2006

[22] ISO 13372. Condition Monitoring and Diagnostics of Machines — Vocabulary; 2012

[23] BS EN 13306. Maintenance. Maintenance Terminology; 2010

[24] Carnero MC. An evaluation system of the setting up of predictive maintenance programmes. Reliability Engineering and System Safety. 2006;**91**(8):945-963

[25] Davis R. An Introduction to Asset Management A Simple but Informative Introduction to the Management of Physical Assets; 2012

[26] Milje R. Engineering Methodology for Selecting Condition Based Maintenance2011. pp. 1-57

[27] Manzini R, Regattieri A, Pham H, Ferrari E. Maintenance for Industrial Systems. London: Springer London; 2010

[28] Narayan V. Effective Maintenance Management: Risk and Reliability Strategies for Optimizing Performance. New York: Industrial Press Inc.; 2004. 128 p. ISBN: 0-8311-3178-0

[29] Langseth H, Haugen K, Sandtorv H. Analysis of OREDA data for maintenance optimisation. Reliability Engineering and System Safety. 1998;**60**(2):103-110

[30] Management SINTEF Industrial. OREDA Offshore Reliability Data Handbook; 2002. p. 835

[31] Tsang AHC. A strategic approach to managing maintenance performance. Journal of Quality in Maintenance Engineering. 1995;**4**(2):87-94. DOI: 10.1108/135525198

[32] Muchiri P, Pintelon L, Gelders L, Martin H. Development of maintenance function performance measurement framework and indicators. International Journal of Production Economics. 2011;**131**(1):295-302

[33] Bendell T. An overview of collection, analysis, and application of reliability data in the process industries. IEEE Transactions on Reliability. 1988;**37**(2):132-137

[34] Zhang W, Jia MP, Zhu L, Yan XA. Comprehensive overview on computational intelligence techniques for machinery condition monitoring and fault diagnosis. Chinese Journal of Mechanical Engineering (English Edition). 2017;**30**(4):782-795

[35] Sikorska JZ, Hodkiewicz M, Ma L. Prognostic modelling options for remaining useful life estimation by industry. Mechanical Systems and Signal Processing. 2011;**25**(5):1803-1836

[36] Crespo Márquez A, Moreu de León P, Gómez Fernández JF, Parra Márquez C, López Campos M. The maintenance management framework. Journal of Quality in Maintenance Engineering. 2009;**15**(2):167-178

[37] ISO 55001. Asset Management — Management systems — Requirements. ISO; 2015

[38] Guevara F, Patiño C, Souza G. "Aplicación del mantenimiento centrado en confiabilidad como herramienta para el incremento de vida operacional de activos mineros," 2012

[39] ISO 9001. Quality Management Systems — Requirements; 2007

[40] Bedoya Rios S, Mesa Roldan CJ, Guevara Carazas FJ. Gestión de mantenimiento y seguridad vial en el marco de la norma UNE-ISO 39001:2015–Caso de estudio Medellín-Colombia; 2017. p. 11

[41] Deming Prize Committee. The Application Guide for The Deming Prize The Deming Grand Prize For Companies and Organizations Overseas; 2015

[42] UNE-EN 16646, Mantenimiento–Mantenimiento en la gestión de los activos físicos; 2015

[43] Carro Paz R, González Gómez D. "Diseño y selección de procesos" [Online]. Available at: http://nulan.mdp.edu.ar/cgi/export/eprint/1613/BibTeX/nulan-eprint-1613.bib [Accessed: 06 December 2018]

[44] J. Bravo Carrasco, Gestion de procesos, 5°. Santiago de Chile, 2013

[45] Palmer RD. Maintenance planning and scheduling handbook. Vol. 912005

[46] Latorella KA, Prabhu PV. Present address: The Eastman Kodak Company, 901 Elm-grove Rd. International Journal of Industrial Ergonomics. 2000;**26**:133-161

[47] Lock MWB, Strutt JE. Reliability in in-service inspection of transport aircraft structures. Civil Aviation Authority CAA Report 85013. London; 1985

[48] Tsang AHC, Yeung WK, Jardine AKS, Leung BPK. Data management for CBM optimization. Journal of Quality in Maintenance Engineering. 2006;**12**(1):37-51

[49] Guevara Carazas FJ, Marthade Souza GF. Fundamentals of maintenance. In: Thermal Power Plant Performance Analysis. 2012

Chapter 2

Advantages of Condition-Based Maintenance over Scheduled Maintenance Using Structural Health Monitoring System

Ting Dong, Raphael T. Haftka and Nam H. Kim

Abstract

This chapter quantifies the advantages of condition-based maintenance on the safety and lifetime cost of an airplane fuselage. The lifecycle of an airplane is modeled as blocks of crack propagation due to pressurization interspersed with inspection and maintenance. The Paris-Erdogan model with uncertain parameters is used to model fatigue crack growth. The fuselage skin is modeled as a hollow cylinder, and an average thickness is calculated to achieve a probability of failure in the order of 1 in 10 million with scheduled maintenance. Condition-based maintenance is found to improve the safety of an airplane over scheduled maintenance and will also lead to savings in lifecycle cost. The main factor of the savings stems from the reduced net revenue lost due to shortened downtime for maintenance. There are also other factors such as work saved on inspection and removing/installing surrounding structures for manual inspection. In addition to cost savings, some potential advantages of condition-based maintenance are discussed such as avoiding damage caused by removing/installing surrounding structures, more predictable maintenance, and improving the safety issues of same aircraft model by posting the frequently occurred damages into Airworthiness Directives, Service Bulletins, or Service Letters.

Keywords: condition-based maintenance, structural health monitoring, damage tolerance, lifecycle cost

1. Introduction

Traditionally, aircraft structures have been designed using the damage tolerance concept (Hoffman, [1], Simpson et al. [2]), which refers to the ability of structure to sustain anticipated loads in the presence of certain damage until such damage is detected through inspections or malfunctions and repaired [3]. More specifically, as cracks on fuselage skin are the damage this chapter is focusing on, it means that structure is designed to withstand small cracks and large cracks are repaired through scheduled inspection and maintenance. In damage tolerance design, an airframe is regularly inspected so that potential damages are early identified and repaired. As such, scheduled maintenance is the primary tool in aircraft maintenance philosophy where inspections and repair works are performed at fixed scheduled intervals in order to maintain a desired level of safety.

IntechOpen

Historically, the risk due to fatigue cracks in fuselage has been identified early in civil aviation due to the three accidents of Comet aircraft (BOAC Flight 783 (1953), BOAC Flight 781 (1954), South African Airways Flight 201 (1954)). In addition, the accident of Aloha Airlines Flight 243 (1988) revealed that multiple-site fatigue cracking caused the failure of the lap joint. Fatigue cracks also caused accidents in other parts of the aircraft, such as the wing spar failure in Northwest Airlines Flight 421 (1948). Since then, inspection and scheduled maintenance have been conducted to detect fatigue cracks and repair them before they cause structural failure. However, deficiency and mishap during the inspection and maintenance often caused accidents. For example, the accident of Aloha Airlines was partly caused by the fact that the inspection was conducted at night. Japan Airlines Flight 123 (1985) crashed due to incorrect splice plate installation during the corrective maintenance, which reduces the part's resistance to fatigue cracking to about 70%.

Scheduled maintenance can be categorized into transit check, 24 h of check, and A/B/C/D checks with increasing intensity and interval. For a Boeing 737-300/400/500, the typical C check is carried out at about 2800 flight cycles (4000 flight hours with an average flight length of 1.4 h) [4]. This inspection schedule is chosen such that the probability of an undetected crack growing beyond the critical size before the next scheduled maintenance is less than 1 in 10 million [5].

In CBM, a damage parameter is continuously monitored by a structural health monitoring (SHM) system, whereby maintenance is requested when the value of damage parameter exceeds a certain threshold [6]. Such an SHM system uses onboard sensors and actuators, enabling the damage assessment to be performed as frequently as needed.

This chapter presents an estimate of cost savings using condition-based maintenance over scheduled maintenance. The effect on cost and safety of condition-based maintenance using SHM system over scheduled maintenance is demonstrated for fuselage skin subject to fatigue crack growth. In scheduled maintenance, maintenance is scheduled at predetermined intervals. Since these inspection intervals are relatively large, all detectable cracks must be repaired. In condition-based maintenance, however, crack assessment can be performed as frequently as needed; repair work is then requested only when the size of detected crack exceeds a certain threshold that can threaten the safety of fuselage skin. This leads to condition-based maintenance using SHM to be an effective approach to reduce lifecycle cost. Boller [7] observed that using SHM for condition-based maintenance would lead to lower downtime and inspection cost. Sandborn and Wilkinson [8] and Scanff et al. [9] studied the cost estimation of electronic and helicopter systems, respectively, using health monitoring systems. In order to facilitate a progressive transition from scheduled maintenance to condition-based maintenance, a hybrid approach is also considered where scheduled maintenance is used for critical structures and condition-based maintenance for noncritical structures.

Several simplifications are made in this chapter in order to make the cost calculation simple:

Firstly, although three types of crack detection approaches have been used in scheduled maintenance, general visual inspection (GVI) is considered as the only detection approach in this chapter because it is the most commonly used inspection method. The three detection approaches are:

- General visual inspection (GVI)
- Detailed visual inspection (DVI)
- Nondestructive test (NDT) with increasing resolution

NDT can be subcategorized into eddy current, ultrasonic, X-ray, magnetic particle, and penetrant [10]. For the most part of fuselage skins, GVI is used. As areas that require DVI and NDT are extremely small compared to those that require GVI, it is assumed that GVI is the only detection approach herein.

Secondly, repair of fuselage skin is considered to be the only maintenance in this chapter. In scheduled maintenance, the maintenance of fuselage skin includes repair and replacement. However, replacement of fuselage skin is only performed when unexpected damage in fuselage skin occurs because of incidents, such as the aircraft bumping into a ground vehicle when taxiing or when widespread fatigue damage occurs on aged aircraft. The latter refers to the simultaneous presence of cracks at multiple locations that are of sufficient size and density resulting in the structure not being able to meet any longer required damage tolerance limits; thus, it will not maintain required residual strength after partial structural failure. Under normal circumstances, for a single crack on fuselage skin, the probability of replacing fuselage skin is extremely low based on the first author's experience and can be negligible. Therefore, this chapter discusses only the repair of fuselage skin.

Lastly, the loading condition for every aircraft structural component is complicated, and variable amplitude loadings and repeated hard landing, for example, should be considered. In this study, however, the discussion is focused on crack propagation on fuselage skin. The most dominant loadings are repeated pressurizations during takeoff and landing. Therefore, the pressurization difference is assumed to be the only loading condition herein.

The structure of the chapter is as follows:

In Section 2, the literature on SHM sensor technologies are reviewed. In Section 3, the processes of damage detection and repair are explained. Section 4 quantifies the parameters for scheduled and condition-based maintenance to maintain a specific level of safety. Section 5 compares the cost savings of condition-based maintenance over scheduled maintenance. Section 6 discusses some potential advantages of condition-based maintenance, followed by conclusions in Section 7.

2. Literature review on structural health monitoring technologies

In CBM, the inspection is performed using sensors installed on the aircraft structure, called a SHM system. Therefore, it is important to review the current sensor technologies to evaluate their performance in detecting cracks. In general, the sensors used in SHM systems are either active or passive sensors. Passive sensors detect signals generated by damage due to the evolution of the damage, which does not require an external excitation. Acoustic emission belongs to this category [11]. If damage is detected during flight, this can be a useful method. As mentioned earlier, however, since the inspection is performed on the ground, it would be difficult to use passive sensors to detect damage. Therefore, passive sensors will not be discussed in this chapter.

Active sensors detect damage by sending a signal to the damage. Since the purpose is to use them for SHM, the review in this section focuses on the smallest size of detectable damage, the detection range, the weight of SHM systems, and the possibility of detecting closed cracks. It would be desirable that the SHM systems can detect at least the same damage size with the NDT. The detection range will determine the total number of sensors required to inspect the entire fuselage panels. In order to reduce the payload loss, it is important to reduce the weight of the SHM system. Since the inspection is performed on the ground, it is required to detect closed cracks.

The most widely used active sensor is the piezoelectric wafer active sensor (PWAS), which uses ultrasonic lamb waves. As an actuator, it converts the electric signal to mechanical motion to generate a longitudinal or transverse wave, which propagates on the panel and is reflected at a crack. As a sensor, it receives a wave reflected from a crack and converts it to electric signals. The location and size of damage are estimated by measuring the time, amplitude, or frequency of the reflected wave.

In general, two methods are used to detect damage [6]. In the pulse-echo method, one PWAS sends waves and receives waves reflected at a crack. In the pitch-catch method, one PWAS sends waves, and the other PWAS receives the waves. In addition, several PWAS, called a phase array, are used simultaneously to improve detection capability [12]. Although the abovementioned two methods require undamaged (pristine) state, the time reversal method [13] does not require it. Since the mechanism of detecting damage using PWAS is similar to conventional NDT ultrasonics, the detectable damage size is also similar to NDT. The most preferable feature of PWAS is its capability of detecting a remote damage from the sensor. Giurgiutiu [14] showed a lamb wave tuning method to detect a remote damage effectively. It has been shown that PWAS can be used for both metallic and composite panels [15]. In order to reduce the excessive number of wires to connect sensors, SMART layer [16] is developed by printing circuits of 30 sensors into a thin dielectric film.

Fiber Bragg grating (FBG) uses a series of parallel lines of optical fiber with different refractive indices [17]. When a local strain is produced due to the presence of a crack, it will change the spacing between gratings, which shifts the wavelength of the reflected wave. FBG sensors detect damage by measuring the shift of reflected wavelength. It is small and lightweight. It was shown that a single optical fiber could incorporate up to 2000 FBG sensors [18]. The literature also showed that it could detect barely visible impact damage in a composite panel [19]. However, FBG sensors have a very short detection range because the local strain diminishes quickly as the distance increases. It would perform better for hotspot damage monitoring, where the damage location is already known. Since cracks in fuselage are opened during flight and closed on the ground, FBG is not appropriate for on-ground SHM. Lastly, since FBG measures the change in strains, it requires strains at the undamaged (pristine) state. If there is pre-existing damage, it can only measure the change from the previous damage.

Comparative vacuum monitoring (CVM) sensors are composed of alternating vacuum and atmospheric pressure galleries and detect cracks using pressure leakage between galleries. The testbed in Sandia National Laboratory showed that CVM could detect cracks in the size of 0.02 in [20]. Airbus [21] and Delta Airlines [22] also tested the feasibility of CVM on SHM. CVM sensors are lightweight made of polymer, and the gallery can be as small as 10 μm [23]. Even if CVM sensors do not require undamaged (pristine) state, it can only detect damage underneath the sensor. Therefore, CVM is appropriate for hotspot monitoring. For fuselage damage monitoring, it would require a sensor layout with a very high density.

There are other kinds of sensors, such as carbon nanotube sensors [24], printed sensors [25], and microelectromechanical systems sensors [26]. These sensors are, however, still in the research or development stage and take more time to be commercially available.

As a summary, among different sensor technologies, it turned out that PWAS is the most appropriate for an SHM system for airplane fuselage monitoring as it can detect cracks that are relatively small and far away from the sensors.

3. Maintenance process for fuselage structures

3.1 Corrective maintenance procedure

Repeated pressurization during takeoff and landing of an airplane can cause existing cracks on a fuselage skin to grow, for example, Aloha Airlines Flight 243. The rate of crack growth is controlled by, among other factors:

- The size of initial cracks due to manufacturing or previous maintenance
- The pressure differential between the cabin and the outside atmosphere
- The thickness of the fuselage skin

If left unattended, the cracks may grow to cause fatigue failure of the fuselage skin. In damage tolerance design, the less frequent the inspection, the lower the damage size threshold for repairing cracks in order to maintain a desired level of safety. The action of repairing cracks on fuselage skin to maintain a desired level of safety until the next scheduled maintenance is termed corrective maintenance. This section explains the modeling of the corrective maintenance procedure undertaken to prevent fatigue failure due to excessive crack growth.

The size of cracks in fuselage structures in a fleet of airplanes is modeled as a random variable characterized by a probability distribution that depends on manufacturing and the loading history of the airplane. The corrective maintenance procedure changes this distribution by repairing large-sized cracks as illustrated in **Figure 1**. **Figure 1** is presented as a probability density function (PDF) versus crack length. The solid curve represents the crack size distribution of an airplane entering the maintenance hangar. Different cracks grow at different rates because of random distribution of the Paris-Erdogan model parameters. The maintenance process is designed to repair fuselage skin with cracks larger than a repair threshold. Since crack detection is not perfect due to inspector's capability [27], maintenance only partially truncates the upper tail of the distribution, as represented by the dashed curve in **Figure 1**. It is noted that while there is uncertainty in damage detection, it is assumed that the size of the detected damage is known without any error/noise.

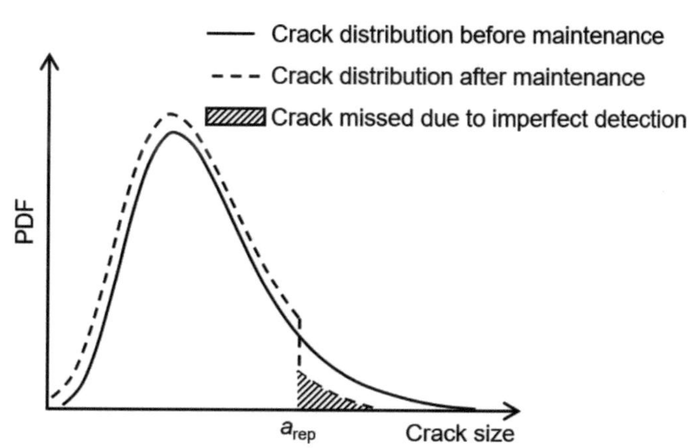

Figure 1.
The effect of inspection and repair process on crack size distribution.

The shaded area represents the fraction of cracks missed during maintenance because of detection imperfection. The cracks that are missed during maintenance and happen to grow beyond the critical crack size before the next maintenance affects the safety of the aircraft.

3.2 Scheduled maintenance

The flowchart in **Figure 2** depicts the scheduled manual maintenance, in which maintenance is programmed at specific predetermined intervals (every N_{man} flight cycles) and corrective action is taken to ensure the airworthiness of the airplane until the next scheduled maintenance.

As all detected cracks on fuselage skins are repaired, the desired level of safety is determined by detection resolution/capability of GVI, a_{gvi}. It is expected that trained inspectors are able to detect cracks larger than 0.5 in (12.7 mm) in GVI. This is also the threshold for repair in scheduled maintenance.

Three parameters affect the lifecycle cost and safety of an aircraft undergoing scheduled maintenance: the maintenance interval, N_{man}; the threshold for repair (detection capability), a_{gvi}; and the thickness of the fuselage skin, t. To achieve a certain desired level of safety, N_{man} and a_{gvi} are correlated with each other. These three parameters together determine the number of maintenance trips and the number of cracks needed to be repaired on fuselage skins.

3.3 Condition-based maintenance

The condition-based maintenance process tracks crack growth continuously and requests maintenance when the crack threatens safety. In this chapter, the condition-based maintenance is considered to be performed using SHM technique. This technique employs onboard sensors and actuators, which are embedded in the structure, to monitor existing crack condition. In doing so, they detect cracks in metallic structures using guided waves transmitted from one location and received

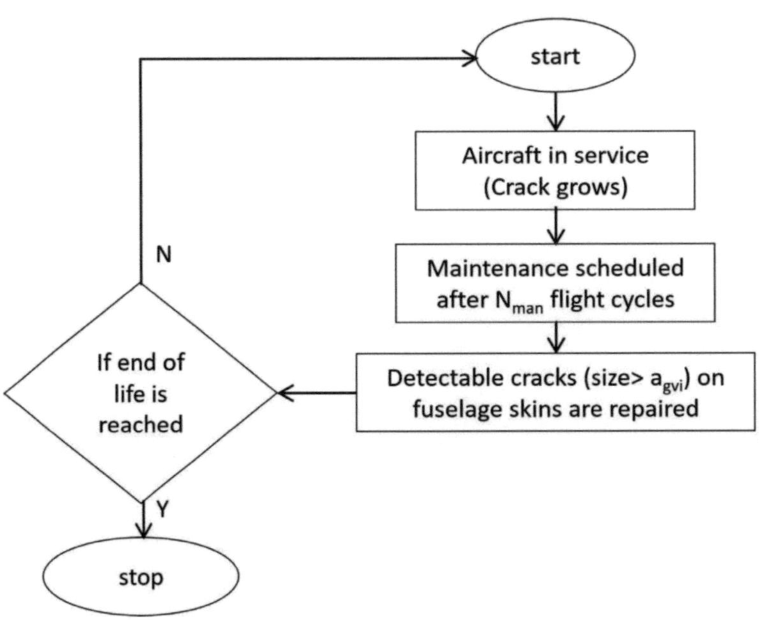

Figure 2.
Flowchart of the scheduled maintenance.

at a different one. The analysis of the change in a guided wave's shape, phase, and amplitude yields indications about crack presence and extension. The probability of detection of the SHM method is comparable with that of conventional ultrasonic and eddy current methods [28]. Crack size and location can be displayed on ground equipment when connecting to onboard sensors and actuators after landing. On-ground equipment can reduce the flying weight and thus may lower the lifecycle fuel cost.

The abovementioned process is called herein maintenance assessment. SHM-based maintenance assessment can be performed as frequently as every flight. However, as the crack increases by only a small amount in each flight cycle, it is unnecessary to perform this assessment after every flight. Also, maintenance assessment is not completely cost-free but requires a small amount of time and personnel. Typically, this assessment frequency (N_{shm}) is assumed to coincide with the A check of scheduled maintenance, which is about 180 flight cycles (250 flight hours with average flight length of 1.4 h [4]).

Figure 3 delineates the condition-based maintenance process. During the assessment, maintenance is requested if the crack size on a fuselage skin exceeds a specified threshold (a_{th}). This threshold is performed, so as to repair all detected cracks on fuselage skins with threatening crack sizes. Additionally, the threshold for threatening crack size $(a_{rep-shm})$ is set substantially lower than the threshold for requesting maintenance (a_{th}) to prevent too-frequent maintenance trips for that airplane.

Condition-based maintenance is controlled by the following parameters:

- The thickness of fuselage skin (t), which affects the crack growth rate
- The thickness (t), along with the frequency of assessment (N_{shm}), and the threshold for requesting maintenance (a_{th}) affect the safety of the airplane
- The threshold for repair $(a_{rep-shm})$ determines the number of cracks needed to be repaired on fuselage skin. It is also set to prevent frequent maintenance trips

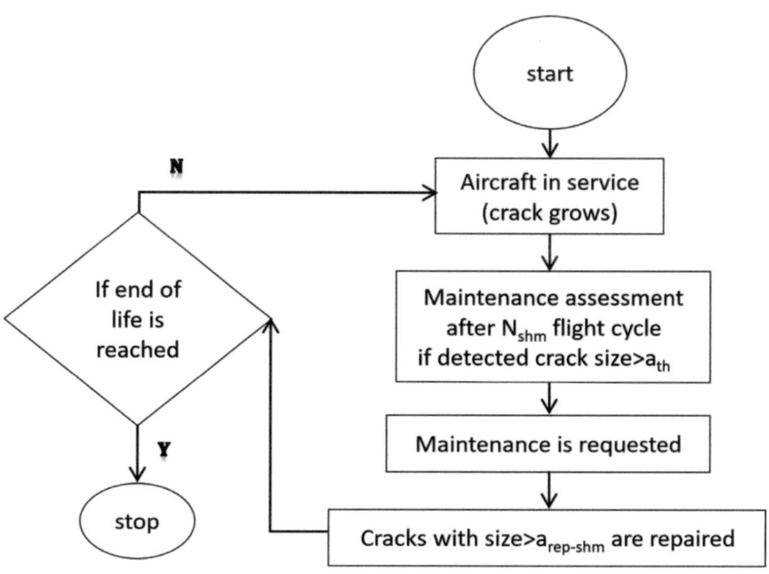

Figure 3.
Flowchart of the condition-based maintenance.

4. Parameters assumed for scheduled and condition-based maintenance

Cracks that are missed or intentionally left unattended during maintenance and grow to critical size before the next maintenance interval affect the safety of the aircraft structure. In the case of scheduled maintenance, the thickness of the fuselage skin (t), the interval of scheduled maintenance (N_{man}), and the threshold for repair (a_{gvi}) affect the aircraft's safety, which is influenced by the thickness of the fuselage skin (t), the frequency of maintenance assessment (N_{shm}), and the threshold for requesting maintenance (a_{th}).

This section deals with quantifying the range of parameters for scheduled and condition-based maintenance. As such, each damage instance is modeled as a through-the-thickness center crack in an infinite plate subject to Mode-I fatigue loading, as shown in Appendix A. The uncertainty in the loading condition and material parameters are summarized in **Table 4**. A crack grows due to pressure differential between the cabin and atmosphere, which is modeled by the Paris-Erdogan model, as shown in Appendix A. From fracture mechanics, the critical crack size (Eq. (3)) to cause failure of a fuselage skin depends on the pressure load and, hence, may also be modeled as a probability distribution. This chapter considers a fuselage skin to be failed if the crack grows undetected beyond the 10^{-7} percentile of critical crack size distribution.

In the scheduled maintenance of a B737-300/400/500, the C check is carried out at about every 2800 flight cycles ($N_{man} = 2,800$) [4] for an airplane life of 50,000 flights. The threshold for repair is equal to the detection capability of GVI, $a_{gvi} = 0.5$ in (12.7 mm). The fraction of cracks which cause failure of fuselage skins due to excessive crack propagation until the end of life is computed by Monte Carlo

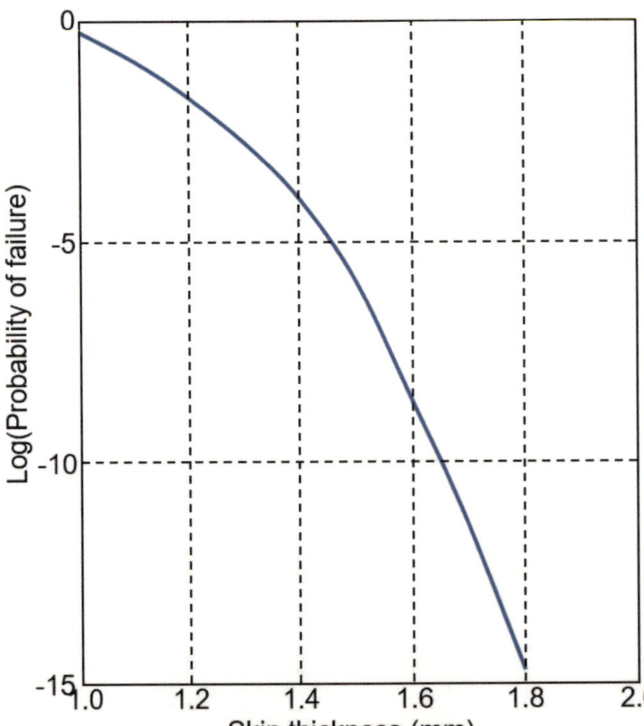

Figure 4.
Variation of lifetime (50,000 flight cycles) probability of failure as a function of fuselage skin thickness for scheduled maintenance at every 2800 flight cycles.

simulations. A fleet of 20,000 airplanes with 500 initial cracks per airplane due to manufacturing or previous maintenance are considered. These cracks are distributed on fuselage skins. The initial crack size and crack growth parameters (m, a_{th}) are randomly sampled for each crack. Pressure is also assumed to vary in each flight. The fraction of cracks that cause fuselage skins to fail is computed for different values of skin thickness, and the variation is plotted in **Figure 4**. Based on **Figure 4**, a fuselage skin with a minimum thickness of 0.06 in (1.53 mm) is required to achieve the target probability of failure of 10^{-7}. Considering that 0.063 in (1.6 mm) is the most common thickness of a typical fuselage skin, this calculation provides a reasonable estimate.

In condition-based maintenance, the threshold for scheduling aircraft to maintenance must be chosen in such a way so as to satisfy the reliability constraint until the next maintenance assessment(N_{shm}). The latter has been chosen as 180 flight cycles, which is equivalent to the current A check interval. If say the threshold for requesting maintenance (a_{th}) is fixed at 1.57 in (40 mm), the reliability for the given value of a_{th} and N_{shm} can be computed using a direct integration procedure, detailed in Appendix C, and is proven to satisfy the desired level of safety.

5. Cost comparison between two maintenance processes

In this chapter, the lifecycle cost of an airplane is considered to be the sum of manufacturing cost, fuel cost incurred during lifecycle, and maintenance cost. Other costs that remain constant for two different approaches are not considered. Cost comparison of two maintenance approaches is discussed in two aspects: cost increase and cost decrease. **Table 1** summarizes the parameters that are used for cost calculation for the two maintenance processes based on Boeing 737-300 Structural Repair Manual and estimated the cost in the maintenance field.

Based on the Structure Repair Manual of a Boeing 737-300, the fuselage skin in the pressurized area is not a regular cylinder. However, it was assumed to be a cylinder to simplify calculation, by using the average diameter $D = 148\text{in}$. In addition, the length of the cylinder can be calculated as $L = 977\text{in}$. As already stated, the thickness of the fuselage skin varies from station to station; however, the most common thickness of $t = 0.063\text{in}$ is used herein. In addition, the density of fuselage skin, which is made of aluminum alloy 2024-T3, is about $\rho = 0.1\text{lb}/\text{in}^3$. Therefore, the total weight of fuselage skin in the pressurized area is $W = \pi DLt\rho = 2957\text{lb}$.

5.1 Cost increased

(1) Manufacturing cost.

Manufacturing cost with SHM system: $600/lb.
Manufacturing cost without SHM system: $500/lb.

Weight of fuselage skins	2957 lb. [10]
Interval of C check	2800 flight cycles
Life cycles	50,000 flight cycles
Net revenue lost due to downtime	$27,000/airplane/day
Labor cost in hangar	$60/h

Table 1.
Parameters for maintenance cost calculation.

$$\text{Cost increased} : (600 - 500) \times 2,957 = 3 \times 10^5 (\$)$$

(2) Cost on replacing SHM equipment.

A finite life of 12,000 flight cycles for SHM equipment is assumed so that the system will need to be replaced four times during 50,000 flight cycles. The lifetime cost for replacing the SHM system after manufacturing is as follows:

$$\text{Cost increased} : 3 \times 10^5 \times 4 = 1.2 \times 10^6 (\$)$$

(3) Fuel cost.

Weight penalty: lifetime fuel consumption cost per aircraft weight. Kaufmann et al. [29] used $1000 per pound as the lifetime fuel cost for 1 pound of gross weight of aircraft. About 5% extra weight is considered for fuselage skin with SHM equipment. Therefore, the cost increase due to SHM equipment weight increased is as follows:

$$\text{Cost increased} : 2957 \times 5\% \times 1000 = 1.5 \times 10^5 (\$)$$

5.2 Cost decreased

As damage assessment intervals in condition-based maintenance are much smaller than that of the scheduled maintenance, the threshold a_{th} for requesting condition-based maintenance to be much larger than a_{gvi} in scheduled maintenance. This high damage tolerance reduces the number of maintenance trips. In addition, because the threshold for repair $a_{rep-shm}$ is larger than a_{gvi}, the number of cracks that are repaired is reduced in condition-based maintenance. It is assumed that these are two factors that would cause savings in aircraft lifecycle maintenance costs.

Monte Carlo simulation (MCS) is performed to compute the number of maintenance trips and the number of cracks repaired on fuselage skins for scheduled and condition-based maintenance. It is assumed that 500 initial cracks on a B733 are distributed on fuselage skins, showing a typical thickness of 0.063 in (1.6 mm).

The damage detection process is governed by the Palmberg expression (Appendix B) with different parameters for scheduled and condition-based maintenance. The parameters computed are listed in **Table 2**. Values in parentheses are MCS standard deviations based on 20,000 airplanes. It is considered that SHM equipment is replaced every 12,000 flight cycles.

It is noted that for the same fuselage skin thickness, condition-based maintenance leads to better reliability and lower number of maintenance trips and cracks repaired. The reason is that scheduled maintenance repairs all the cracks that might grow to threaten safety until the next maintenance, while the condition-based maintenance repairs only those that actually grow to threaten safety.

Based on the results computed above, the cost saved can be calculated as follows:

(1) Net revenue saved due to shortened downtime.

Types of maintenance	Probability of failure	Avg. no. of maintenance trips per airplane	Avg. no. of cracks repaired/airplane
Scheduled	1E-8	18	10 (0.6)
Condition-based	1E-13	7.6 (0.3)	5.8 (0.2)

Table 2.
Comparison between scheduled and condition-based maintenance.

The downtime for C checks of B737 CL varies from several days to 2 months as the age of the aircraft increases. This chapter regards 30 days as a typical downtime for a C check. Usually, the inspection procedure takes up about 1/3–1/4 of the whole downtime in scheduled maintenance. In condition-based maintenance, however, it is assumed that the assessment process can be completed in 1 day using the SHM system. Therefore, about 7 days can be saved on inspection.

Downtime is shortened not only because of the efficient assessment process in condition-based maintenance but also due to the time spent on removing/installing the surrounding structures for GVI in scheduled maintenance. In the latter case, the general visual inspection can only be carried out when surrounding structures are removed. For example, if general visual inspection is performed on fuselage skins in the cargo area, all floor panels, sidewalls, insulation blankets, etc. have to be removed. Downtime of CBM can be reduced by about 5 days by skipping this procedure.

From the analysis above, the downtime can be shortened by 12 days for each maintenance trip in condition-based maintenance. Therefore, the downtime for condition-based maintenance is assumed as 18 days for each maintenance trip:

$$\text{Cost saved}: 27,428 \times (18 \times 30 - 7.8 \times 18) = 1.1 \times 10^7 (\$)$$

(2) Inspection cost.

As stated above, the time shortened on inspection by using SHM system is 7 days. Assume that 100 h of labor is needed on inspection per day at $60/h:

$$\text{Cost saved}: 7 \times 100 \times 60 \times 18 = 7.56 \times 10^5 (\$)$$

(3) Cost for removing/installing surrounding structures.

The time spent on removing/installing surrounding structures for easy access of GVI is about 5 days with 300 h of labor per day:

$$\text{Cost saved}: 5 \times 300 \times 60 \times 18 = 1.62 \times 10^6 (\$)$$

(4) Crack repair cost.

As calculated above, the number of cracks that need to be repaired is 10 in scheduled maintenance and 5.8 in condition-based maintenance for each maintenance trip. Fuselage skin with cracks detected is repaired by different methods depending on the size of the crack [10]. In the case of fuselage skin, the doubler repair is the most common method. Although different repair methods are adopted according to the size of the crack, in this chapter, it is assumed that the typical doubler repair be implemented.

For a doubler of 10 × 10 in, 60 h of labor is needed. The cost for this doubler repair is about $360 with $60 labor cost per hour:

$$\text{Cost saved}: 360 \times (18 \times 10 - 7.6 \times 5.8) = 4.9 \times 10^4 (\$)$$

Table 3 summarizes cost increase and decrease for the two maintenance strategies. It can be concluded from the table that total cost saved is about $\$1.18 \times 10^7$, which is about 10% of the lifecycle cost, by using SHM system on condition-based maintenance over scheduled maintenance. The main factor leading to this cost savings is the reduced net revenue lost due to shortened downtime. The effect of cost saved on inspection and removing/installing the surrounding structures is

Cost increased ($)	Manufacturing cost	SHM replacement		Fuel cost	Total
	3×10^5	1.2×10^6		1.5×10^5	1.65×10^6
Cost decreased ($)	Net revenue saved	Inspection cost	Removing/installing cost	Crack repair cost	Total
	1.1×10^7	7.56×10^5	1.62×10^6	4.9×10^4	1.34×10^7
Total cost saved ($)		1.18×10^7			

Table 3.
Summary of cost increased and decreased for two maintenance approaches.

relatively small (20% of the total cost saved). It is also noted that cost saved by the reduced number of cracks repaired is negligible.

6. Potential advantages of condition-based maintenance

In addition to the cost savings calculated, some further potential benefits may be gained by using SHM system on condition-based maintenance. Firstly, skipping the removing/installing surrounding structure procedures in SHM systems not only saves time and labor but also prevents potential damage to structures caused by the removing/installing process. Although the fasteners can be replaced after each removing/installing, the fastener holes, taking rivet holes, for example, will be enlarged after each repair work. This is an irreversible damage and might also be the source of new cracks. Even worse, some accidents might occur during the removing/installing, such as drilling through unrelated structures. All these are troublesome issues reported by MRO companies and airlines frequently. With the introduction of an SHM system, this problem may be eliminated.

Secondly, maintenance is more predictable with the SHM system. In scheduled maintenance, damages are detected by manual inspection in the hangar. For some unexpected damages, several days are wasted on preparing special equipment, tools, and/or materials. Sometimes, it even takes a week or so to confirm a repair plan by consulting the manufacturer of the aircraft.

In condition-based maintenance, however, by monitoring the cracks continuously using the SHM system combined with the Paris-Erdogan model and the MCS to model the growth, crack growth and size are more predictable, thus stepping up maintenance and repair work.

Furthermore, with the ongoing research on sensors and actuators, its detection ability will not only be confined on cracks; it can also be used for detecting other typical structural damages such as corrosion, dents, holes, delamination, etc. By collecting and analyzing all the data from the SHM system, the structures on which certain damage frequently occurred affecting the safety of aircraft could be found. These can be posted in Airworthiness Directives (AD), Service Bulletins (SB), or Service Letters (SL), to help eliminate the potential safety issues in the whole fleet of same aircraft model.

7. Conclusions

Two maintenance approaches are discussed in this chapter. Traditionally, scheduled maintenance is carried out at predetermined intervals to maintain a desired level of safety. Recently, with the development of SHM techniques, condition-based maintenance uses onboard SHM sensors and actuators to detect damage on fuselage skins, which, in turn, may be performed as frequently as needed. Hence, maintenance is requested only when a particular condition is met.

The improved reliability and cost savings of condition-based maintenance over scheduled one are discussed. As the usage of onboard SHM system, downtime for each maintenance trip is shortened significantly in condition-based maintenance, leading to considerable cost saving of net revenue. This SHM system also avoids removing/installing the surrounding structures. All these factors may lead to significant cost savings in CBM. In addition, some potential advantages of condition-based maintenance are discussed in this chapter, which includes reducing the possibility of human error during the maintenance process, preparing maintenance equipment in advance, and using the same sensors to detect other types of damages.

Acknowledgements

This research was partly supported by NASA Langley Research Center (Contract No. NNX08AC33A). The authors gratefully acknowledge this support.

List of abbreviations

CBM	condition-based maintenance
CVM	comparative vacuum monitoring
DVI	detailed visual inspection
FBG	fiber Bragg grating
GVI	general visual inspection
MCS	Monte Carlo simulation
MRO	maintenance, repair, and overhaul
NDT	nondestructive test
PDF	probability density function
PWAS	piezoelectric wafer active sensor
SHM	structural health monitoring

Appendices

A. Fatigue damage growth due to fuselage pressurization

Fatigue crack growth can be modeled in a number of ways. Beden et al. [30] provided an extensive review of crack growth models. Mohanty et al. [31] used an exponential model to model fatigue crack growth. Scarf [32] advocated the use of simple models, when the objective was to demonstrate the predictability of crack growth. In this chapter, a simple Paris-Erdogan model [33] is considered to describe the crack growth behavior. However, other advanced models can also be used.

Damage in the fuselage skin of an airplane is modeled as a through-the-thickness center crack in an infinite plate. The life of an airplane can be viewed as consisting of damage growth cycles, interspersed with inspection and repair. The cycles of pressure difference between the interior and the exterior of the cabin during each flight is instrumental in fatigue damage growth. The crack growth behavior is modeled using the Paris-Erdogan model, which gives the rate of damage size growth as a function of half damage size (a), pressure differential (p), thickness of fuselage skin (t), fuselage radius (r), and Paris-Erdogan model parameters, C and m:

$$\frac{da}{dx} = C(\Delta K)^m \tag{1}$$

where the range of stress intensity factor is approximated with the stress $\Delta\sigma = pr/t$ as

$$\Delta K = \Delta\sigma\sqrt{\pi a} \tag{2}$$

The following critical crack size can cause failure of the panel and is approximated as

$$\sqrt{a_{cr}} = \frac{K_{IC}}{\Delta\sigma\sqrt{\pi}} \tag{3}$$

where K_{IC} is the fracture toughness of an infinite plate with a through-the-thickness center crack loaded in the Mode-I direction.

In the above damage growth process, the following uncertainty is considered: uncertainty in the Paris-Erdogan model parameters, pressure differential, and initial crack size. The damage size after N flight cycles depends on the aforementioned parameters and is also uncertain. The values of uncertain parameters are tabulated in **Table 4**.

It is approximated that all fuselage skins are made of aluminum alloy 2024-T3 with dimensions of $57' \times 57' \times 0.063$ (17.4 m × 17.4 m × 1.6 mm). Newmann et al. (Pg 113, **Figure 3**) [34] showed the experimental data plot between the damage growth rate and the intercept and slope, respectively, of the region corresponding to stable damage growth. As the region of the stable damage growth can be bounded by a parallelogram, the estimates of the bounds of the parameters, C *and* m, are obtained from **Figure 3** of Newmann et al. [34].

For a given value of intercept C, there is only a range of slope (m) permissible in the estimated parallelogram. To parameterize the bounds, the left and right edges of the parallelogram were discretized by uniformly distributed points. Each point on

Parameter	Type	Value
Initial crack size (a_0)	Random	LN(0.2, 0.07)mm
Pressure (p)	Random	LN(0.06, 0.003)MPa
Radius of fuselage (r)	Deterministic	2 m (76.5 in)
Thickness of fuselage skin (t)	Deterministic	1.6 mm (0.063 in)
Mode-I fracture toughness (K_{IC})	Deterministic	$36.58 MPa\sqrt{m}$
Paris-Erdogan law constant (C)	Random	U[$\log_{10}$(5E-11), $\log_{10}$(5E-10)]
Paris-Erdogan law exponent (m)	Random	U[3, 4.3]
Palmberg parameter for scheduled maintenance (a_{h-man})	Deterministic	12.7 mm (0.5 in)
Palmberg parameter for scheduled maintenance (β_{man})	Deterministic	0.5
Palmberg parameter for SHM based inspection (a_{h-shm})	Deterministic	5 mm (0.2 in)
Palmberg parameter for SHM based inspection (β_{shm})	Deterministic	5.0

Table 4.
Parameters for crack growth and inspection.

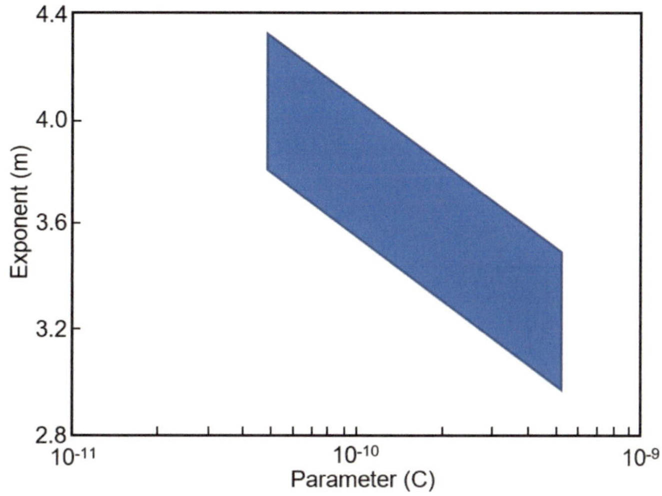

Figure 5.
Possible region of Paris-Erdogan model parameters.

the left and right corresponds to a value of C. For a given value of C, there are only certain possible values of the slope, m. **Figure 5** plots those permissible ranges of slop (m), for a given value of intercept (C). It can be seen from **Figure 5** that the slope and $\log(C)$ are negatively correlated; the correlation coefficient is found to be about -0.8.

B. Inspection model

Kim et al. [35], Packman et al. [36], Berens and Hovey [37], Madsen et al. [38], Mori and Ellingwood [39], and Chung et al. [40] have modeled the damage detection probability as a function of damage size. In this chapter, the inspection of fuselage skins for damage is modeled using the Palmberg equation.

In scheduled maintenance and in SHM-based maintenance assessment, the detection probability can be modeled using the Palmberg Equation [41] given by

$$P_d(a) = \frac{\left(\frac{a}{a_h}\right)^{\beta}}{1 + \left(\frac{a}{a_h}\right)^{\beta}} \tag{4}$$

The expression gives the probability of detecting damage with size $2a$. In Eq. (4), a_h is the half damage size corresponding to 50% probability of detection, and β is the randomness parameter. Parameter a_h represents average capability of the inspection method, while β represents the variability in the process. Different values of the parameter, a_h and β, are considered to model the inspection for scheduled maintenance and also for SHM-based maintenance assessment. **Table 4** shows the parameters used in the damage growth model, as well as the inspection model.

C. Direct integration procedure

The direct integration procedure is a method used to compute the probability of an output variable with random input variables. In general, Monte Carlo simulation can be used to calculate the probability, but it requires many samples, and the results have sampling error. In this chapter, the direct integration process is used to compute the probability of having a specific crack size. The damage size distribution is a function of initial crack size, pressure differential, and Paris-Erdogan model parameter (C, m), which are all random:

$$f_N(a) = h(a_0, f(p), J(C, m)) \tag{5}$$

where $a_0, f_N(a)$, and $f(p)$ represent the initial crack size, the probability density function of crack size after N cycles, and the pressure differential, respectively. $J(C, m)$ is the joint probability density of the Paris-Erdogan model parameters (C, m). The probability of crack size being less than a_N after N cycles is the integration of the joint probability density of input parameters over the region that results in a crack size being less than or equal to a_N, that is,

$$\Pr(a \le a_N) = \int \dots \int_R a_0 J(C, m) f(p) \mathrm{d}R \tag{6}$$

where R represents the region of (a_0, C, m, p) which will give $a \le a_N$.

Based on preliminary analysis performed by the authors, the effect of random pressure differential was averaged out over a large number of flight cycles. Therefore, the average of the pressure differential is used in the following calculation. Hence, Eq. (6) reduces to be a function of m and C, as

$$F_N(40) = \iint_A J(C, m) \, \mathrm{d}C\mathrm{d}m \tag{7}$$

where A represents the region of $\{C, m\}$ that would give $a_N \le 40mm$ for a given initial crack size, a_0. The parallelogram in **Figure 6** is the region of all possible combinations of Paris-Erdogan model parameters, $\{C, m\}$. For the initial crack size, $a_0 = 1mm$, cracks in the gray triangular region will grow beyond 40 mm after $N = 50,000$ cycles. If the initial crack size is distributed, then the integrand is evaluated at different values in the range of the initial crack size, and the trapezoidal rule is used to compute the probability at the desired crack size.

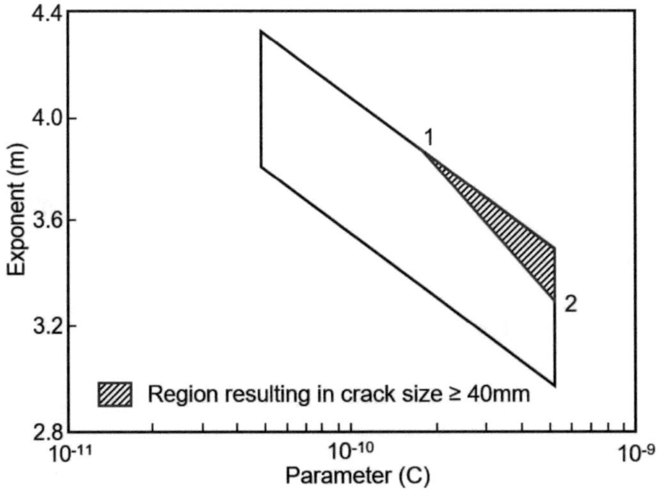

Figure 6.
Regions of $\{C, m\}$ *for* $N = 50,000$ *and* $a_0 = 1mm$.

Author details

Ting Dong, Raphael T. Haftka and Nam H. Kim*
University of Florida, Gainesville, FL, USA

*Address all correspondence to: nkim@ufl.edu

References

[1] Hoffman PC. Fleet management issues and technology needs. International Journal of Fatigue. 2009; **31**:1631-1637

[2] Simpson DL, Brooks CL. Tailoring the structural integrity process to meet the challenges of aging aircraft. International Journal of Fatigue. 1999;**1**: 1-14

[3] Ulf G. Goranson Damage Tolerance Facts and Fiction. In: International Conference on Damage Tolerance of Aircraft Structures. Vol. 5. 2007

[4] Boeing 737-300/400/500 Maintenance Planning Data. D6-38278; 1.0-12

[5] Pattabhiraman S, Gogu C, Kim NH, Haftka RT, Bes C. Skipping unnecessary structural airframe maintenance using on-board structural health monitoring system. Journal of Risk and Reliability. 2012;**226**(5):549-560

[6] Giurgiutiu V, Cuc A. Embedded non-destructive evaluation for structural health monitoring, damage detection, and failure prevention. Shock & Vibration Digest. 2005;**37**(2):92

[7] Boller C. Next generation structural health monitoring and its integration into aircraft design. International Journal of Systems Science. 2000;**31**(11): 1333-1349

[8] Sandborn PA, Wilkinson C. A maintenance planning and business case development model for the application of prognostics and health management (PHM) to electronic systems. Microelectronics Reliability. 2007; **47**(12):1889-1901

[9] Scanff E, Feldman KL, Ghelam S, Sandborn P, Glade M, Foucher B. Life cycle cost estimation of using prognostic health management (PHM) for helicopter avionics. Microelectronics Reliability. 2007;**47**(12):1857-1864

[10] Boeing 737-300 Structural Repair Manual. 51-00-06; 3-4

[11] Bhuiyan MY, Bao J, Poddar B, Giurgiutiu V. Toward identifying crack-length-related resonances in acoustic emission waveforms for structural health monitoring applications. Structural Health Monitoring. 2018;**17**: 577-585

[12] Giurgiutiu V, Bao J. Embedded-ultrasonics structural radar for in situ structural health monitoring of thin-wall structures. Structural Health Monitoring. 2004;**3**:121-140

[13] Xu B, Giurgiutiu V. Single mode tuning effects on Lamb wave time reversal with piezoelectric wafer active sensors for structural health monitoring. Journal of Nondestructive Evaluation. 2007;**26**:123-134

[14] Giurgiutiu V. Tuned Lamb wave excitation and detection with piezoelectric wafer active sensors for structural health monitoring. Journal of Intelligent Material Systems and Structures. 2005;**16**:291-305

[15] Zhao X, Gao H, Zhang G, Ayhan B, Yan F, Kwan C, et al. Active health monitoring of an aircraft wing with embedded piezoelectric sensor/actuator network: I. Defect detection, localization and growth monitoring. Smart Materials and Structures. 2007; **16**:1208

[16] Lin M, Qing X, Kumar A, Beard SJ. Smart layer and smart suitcase for structural health monitoring applications. In: Proceedings of the Smart Structures and Materials 2001: Industrial and Commercial Applications of Smart Structures Technologies, Newport Beach, CA, 14 June 2001.

Vol. 4332. Bellingham, WA, USA: International Society for Optics and Photonics; 2001. pp. 98-107

[17] Staszewski W, Boller C, Tomlinson GR, editors. Health Monitoring of Aerospace Structures: Smart Sensor Technologies and Signal Processing. Hoboken, NJ, USA: John Wiley & Sons; 2004

[18] Di Sante R. Fibre optic sensors for structural health monitoring of aircraft composite structures: Recent advances and applications. Sensors. 2015;**15**: 18666-18713

[19] Takeda S, Aoki Y, Ishikawa T, Takeda N, Kikukawa H. Structural health monitoring of composite wing structure during durability test. Composite Structures. 2007;**79**:133-139

[20] Roach D. Real time crack detection using mountable comparative vacuum monitoring sensors. Smart Structures and Systems. 2009;**5**:317-328

[21] Stehmeier H, Speckmann H. Comparative vacuum monitoring (CVM). In: Proceedings of the 2nd European Workshop on Structural Health Monitoring, Munich, Germany. 2004

[22] Roach DP, Rice TM, Neidigk S, Piotrowski D, Linn J. Establishing the Reliability of SHM Systems through the Extrapolation of NDI Probability of Detection Principles; No. SAND2015-4452C. Albuquerque, NM, USA: Sandia National Laboratories (SNL-NM); 2015

[23] Wishaw M, Barton DP. Comparative vacuum monitoring: A new method of in-situ real-time crack detection and monitoring. In: Proceedings of the 10th Asia-Pacific Conference on Nondestructive Testing, Brisbane, Australia. 2001

[24] Kang I, Schulz MJ, Kim JH, Shanov V, Shi D. A carbon nanotube strain sensor for structural health monitoring. Smart Materials and Structures. 2006; **15**:737

[25] Zhang Y, Anderson N, Bland S, Nutt S, Jursich G, Joshi S. All-printed strain sensors: Building blocks of the aircraft structural health monitoring system. Sensors and Actuators, A: Physical. 2017;**253**:165-172

[26] Varadan VK, Varadan VV. Microsensors, microelectromechanical systems (MEMS), and electronics for smart structures and systems. Smart Materials and Structures. 2000; **9**:953

[27] Good GW, Nakagawara VB & Center MMA. Vision Standards and Testing Requirements for Nondestructive Inspection (NDI) and Testing (NDT) Personnel and Visual Inspectors. Washington, DC: Federal Aviation Administration; 2003

[28] Ihn JB, Chang FK. Pitch-catch active sensing methods in structural health monitoring for aircraft structures. Structural Health Monitoring. 2008; 7(1):5-19

[29] Kaufmann M, Zenkert D, Mattei C. Cost optimization of composite aerospace structures. Composite Structures. 2002;**57**(1):141-148

[30] Beden SM, Abdullah S, Ariffin AK. Review of fatigue crack propagation models for metallic components. European Journal of Scientific Research. 2009;**28**(3):364-397

[31] Mohanty JR, Verma BB, Ray PK. Prediction of fatigue crack growth and residual life using an exponential model: Part II (mode-I overload induced retardation). International Journal of Fatigue. 2009;**31**:425-432

[32] Scarf P. On the application of mathematical models in maintenance.

European Journal of Operational Research. 1997;**99**:493-506

[33] Paris PC, Erdogan F. A critical analysis of crack propagation laws. Journal of Basic Engineering. 1960;**85**: 528-534

[34] Newman JC Jr, Phillips EP, Swain MH. Fatigue-life prediction methodology using small-crack theory. International Journal of Fatigue. 1999; **21**:109-119

[35] Kim S, Frangopol DM. Optimum inspection planning for minimizing fatigue damage detection delay of ship hull structures. International Journal of Fatigue. 2011;**33**:448-459

[36] Packman PF, Pearson HS, Owens JS, Yong G. Definition of fatigue cracks through nondestructive testing. Journal of Materials. 1969;**4**:666-700

[37] Berens AP, Hovey PW. Evaluation of NDE reliability characterization. In: AFWALTR-81-4160. Vol. 1. Dayton, Ohio: Air Force Wright Aeronautical Laboratory, Wright-Patterson Air Force Base; 1981

[38] Madsen HO, Torhaug R, Cramer EH. Probability-based cost benefit analysis of fatigue design, inspection and maintenance. In: Proceedings of the Marine Structural Inspection, Maintenance and Monitoring Symosium1991, SSC/SNAME, Arlington, VA. pp. 1-12

[39] Mori Y, Ellingwood BR. Maintaining reliability of concrete structures: Role of inspection/repair. Journal of Structural Engineering, ASCE. 1994;**120**(3): 824-845

[40] Chung H-Y, Manuel FKH. Optimal inspection scheduling of steel bridges using nondestructive testing techniques. Journal of Bridge Engineering - ASCE. 2006;**11**(3):305-319

[41] Palmberg B, Blom AF, Eggwertz S. Probabilistic damage tolerance analysis of aircraft structures. In: Sih GC, Provan JW. editors. Probabilistic Fracture Mechanics and Reliability. Netherlands: Springer; 1987. pp. 47-130

Chapter 3

Reliability Technology Based on Meta-Action for CNC Machine Tool

Yan Ran, Wei Zhang, Zongyi Mu and Genbao Zhang

Abstract

Computer numerical control (CNC) machines are a category of machining tools that are computer driven and controlled, and are as such, complicated in nature and function. Hence, analyzing and controlling a CNC machine's overall reliability may be difficult. The traditional approach is to decompose the major system into its subcomponents or parts. This, however, is regarded as not being an accurate method for a CNC machine tool, since it encompasses a dynamic working process. This chapter proposes a meta-action unit (MU) as the basic analysis and control unit, the resulting combined motion effect of which is believed to optimize the CNC's overall function and performance by improving each meta-action's reliability. An overview of reliability technology based on meta-action is introduced.

Keywords: reliability, meta-action, CNC machine tool

1. Introduction

Along with social development, the reliability of computer numerical control (CNC) machine tools is becoming more and more important in the market [1]. However, it seems that reliability analysis becomes increasingly difficult, not least due to its complex structure. In order to improve the reliability of CNC machine tools, many scholars have carried out extensive research, including reliability prediction, allocation, analysis, test, and evaluation. There are a series of mature quality technique tools, such as failure mode and effects analysis (FMEA) and fault tree analysis (FTA), to name but a few. Yet, most of these tools are based on the reliability technology of electronic products. The reliability block diagram and the mathematical modeling of the parts are established straightforwardly. In this field, the electronic components, such as resistors and capacitors, do not interact with each other. When assigning a reliability index, the reliability index of the whole machine is allocated to each component according to the reliability block diagram. Then, an FMEA analysis is performed so as to identify all possible failure modes according to historical data and tests [2, 3]. At the end, the FTA analysis of each failure mode is executed in order to determine all bottom events [4]. As such, the reliability of the entire machine is predicted by the component level reliability block diagram.

In reliability research, reliability data is fundamental. The data pertaining to CNC machine tool reliability are not enough, suggesting that the analysis results and accuracy are unsatisfactory [5]. In order to obtain reliability analysis technology suitable for CNC machine tools, various kinds of CNC machine tools were analyzed and summarized. Then, the most basic structure to determine the reliability of CNC machine tool—meta-action was established. In this chapter, this method is standardized.

The FMA decomposition method is described in detail, which is to obtain meta-action. The definition of meta-action and its parts are discussed. The conceptual, structural, and assembly models of meta-action are defined. Identifying similarities of various CNC machine tools may prove difficult, as is their respective analysis. The specific movement and function of each meta-action unit is different, hence, establishing a standardized meta-action model may equally be difficult. According to the meta-action decomposition analysis method, the most important motion unit of CNC machine tool is found, which is affecting its reliability in most of the cases.

This chapter introduces the basic methodology. A number of industrial applications are also presented. The method applies in reliability modeling, allocation, evaluation, and fault diagnosis. Afterward, the research on reliability test and design based on meta-action would be performed. This includes setting up a reliability test bench and performing a meta-action reliability test used in design. All reliability studies may use this method and a complete reliability research system will form. Likewise, this method can be used in other quality characteristics analysis, such as precision, availability, and stability. Thus, further research aimed in this very specific area is deemed necessary.

Indeed, Karyagina proposed that the CNC machine tool manufacturers should pay more attention to the fault information feedback and reliability analysis of after-sales products and to establish a quality and a reliability assurance system [6]. Su and Xu performed research on the theory and methods of dynamic reliability modeling for complex electromechanical products [7]. Zhang and Wang focused on reliability allocation technology of CNC machine tools based on task [8].

Building on past experiences, when analyzing the reliability of such complex systems, the approach would be to break it down into small systems or basic units, and then analyze the basic units instead. There are a number of ways to further divide the machine tool. Xin and Xu took the machining process as the basic unit in precision analysis [9]. Zhang et al. considered the part as the basic unit in the assembly process [10]. Each decomposition method has its own clear object, but few can analyze a system with much function and quality coupling synthetically.

2. Findings

The difference between a CNC machine tool and an electronic product is that the function of the CNC machine tool is realized in terms of the relative motion between components. The latter are internally driven by a large number of meta-actions. Therefore, for a CNC machine tool, as long as the meta-actions break down, the movement function and performance of the components cannot be realized normally. Thus, action should be taken at the basic unit of design, analysis, test, and control [11]. The correctness of each action should be guaranteed to ensure the entire machine's function. All parts that realize an action may be treated as a whole. The method based on action simplifies the analysis process and refines its results. It can be used not only in reliability, but also in the design and manufacture of the CNC machine tool [12]. A complete new theoretical system of CNC machine tool design and manufacture based on meta-action is proposed.

Reliability is the product's ability to perform its specified functions under the stated conditions for a given period of time [13] and is concerned whether the function and movement can be realized. However, the traditional decomposition methods, assembly unit-component-parts (ACP), function-behavior-structure (FBS), and components-suite-parts (CSP), are based on product structure (or parts), which cannot reflect the motion characteristics of a CNC machine tool. Therefore, these methods are not entirely suitable for reliability analyses of dynamic systems.

A CNC machine may perform a number of functions such as drilling and milling throughout its entire service life. Basically, the function is realized by some movements of mechanisms, such as the rotational movement of a spindle. Finally, the movement is gradually achieved by the transmission of basic meta-actions. That is, the function of the CNC machine tool is accomplished by the movement, which in turn is completed by different actions. The latter defines the reliability of the product.

The main reasons that traditional methods are not applicable to CNC machine tool are described as follows:

1. CNC machine tools are basically unable to carry out accurate reliability prediction analysis, as they lack failure-specific probability data. Collecting the data requires much time and cost. In order to obtain more data, many scholars expand the data by means of some mathematical methods. Jia et al. proposed a method of increasing reliability data of CNC machine tool based on artificial neural network theory and algorithm [14]. The radial basis function (RBF) neural network is used to simulate the reliability data, which enlarges the latter's sample size [15]. This method can expand the data; however, the data is not precise.

2. The function and performance of CNC machine tools mainly rely on the interaction between components. It is necessary to analyze these parts as a whole [16].

3. The components of a CNC machine tool are very complex. A component may contain thousands of parts, and the establishment of a fault tree is very large. Zhai used fuzzy methods to solve the minimum cut-set. The large complex fault tree is decomposed into relatively independent sub-trees [17]. However, the basic problem is not resolved.

4. The failure mode of a CNC machine tool is higher than that of an electronic product and as such, the failure reasons may be extensive. Thus, it is difficult to predict all the potential failure modes in FMEA analysis [18].

5. Because of the complexity and the cross fusion of components, the reliability allocation method of electronic products cannot be used directly [19].

The meta-action decomposition method is proposed in this chapter. As such, the CNC machine tool is decomposed into several MUs, which are composed of several parts.

3. Analysis

In this chapter, the meta-action method is described in detail, including the FMA decomposition method and meta-action structure. Some applications of this method

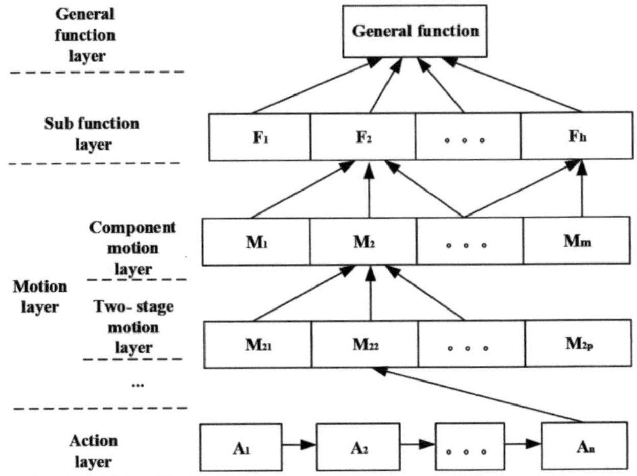

Figure 1.
FMA-structured decomposition.

are represented as well. Also, the use of meta-action reliability technology is described in this section, to provide references for increasing CNC machine tools' degree of reliability. The applications are introduced in three aspects: reliability modeling, design, and manufacturing.

3.1 FMA decomposition method

The "FMA" structured decomposition is used to decompose the CNC machine tool to the meta-action level and carry out the reliability analysis at the MU level. One may conduct the "FMA"-structured decomposition as shown in **Figure 1**. The concrete steps of the meta-action decomposition method are described as follows:

1. Analyze all functions of the CNC machine tool by means of its design project description or instruction manual.

2. According to the structure of the CNC machine tool, study the pattern to realize the function and determine the movements facilitating certain functions

3. Analyze the transfer route from the power part to the actuator(s) and obtain the meta-actions.

4. Depict the FMA tree including functions, movements, and meta-actions based on the above three steps

5. Determine the elements to realize the meta-action and describe the MUs

    - Function layer: it is the design function of the CNC machine tool at the level of design, i.e., milling, grinding, drilling, and turning functions.

    - Motion layer: in order to ensure the normal implementation of a function, the required motion combination level is the motion layer. For example, in order to realize the function of "drilling" in a machining center, it is necessary to co-ordinate the movement of spindle rotation, the indexing of NC turntable, X-Y-Z axes, which form the motion layer

- Action layer: in order to ensure the normal completion of the movement, the level of actions combination is the action layer. The actions in the layer are all meta-actions, and there is no inclusion relationship.

According to the path of action transmission, the actions can be divided into first-order action, second-order action, ..., and N-level action. For example, the movement of the worm and gear drive is divided into two meta-movements. The worm rotation is a first-order action, and the worm gear rotation is a second-order action.

3.2 Meta-action and meta-action unit

3.2.1 Meta-action

The function of a CNC machine tool is accomplished by motion, which in turn is usually done by a transmission system. The latter can be decomposed into the most basic motion unit. Therefore, meta-action may be defined the smallest motion in mechanical products.

The meta-action of CNC machine tool can be usually divided into moving meta-action and rotating meta-action. The former realizes the most basic moving functions, such as the linear movement of piston in the cylinder, the linear movement of a nut along the axis of a screw, the movement of a moving guide rail on the static guide rail, etc.

The latter accomplishes the most basic rotating functions, such as a pair of gear transmission that may be divided into two gear rotating meta-actions. A pair of worm and gear transmission can be decomposed into worm rotating meta-action and worm wheel rotating meta-action.

In design and manufacture, the performance of a CNC machine tool can be controlled only by managing the performance of meta-action.

3.2.2 Meta-action unit

In order to realize the movement of components, the following four elements must be present:

- power input parts,
- transmission parts,
- supporting parts,
- motion output parts.

For example, in order to carry out the rotational motion function of the spindle, it is necessary to have a motor coupling, a pulley, or a gear as the power input, an intermediate drive shaft and a gear as the transmission parts, a supporting part (such as a spindle box) for mounting the transmission parts, and a spindle body as the motion output. These parts form an assembly that facilitates the rotation movement function of the main shaft.

In view of the above, one might define the MU as the unified whole of all parts, which can ensure the normal operation of the meta-action according to the structural relations. The MU must have the following basic elements: power input, power output, middleware, support, and fastener. The specific definition of each basic element is shown in **Table 1**.

3.2.3 The basic model of MU

1. The conceptual model of MU

In order to describe the concept of MU, a conceptual MU model is established, which is shown in **Figure 2**.

2. The structural model of a MU

The structural model describes the MU from the aspect of mechanical structure. In general, two types of MUs units are moving units and rotating units. **Figure 3** is

Composed element	Definition	Example
Power input parts	In MU, the parts that receive or provide a power source or adjacent to the motion or power input of the previous MU	In worm and worm gear drive, the motor is the input of the element action unit, which provides the power input for the element action unit
Power output parts	The last part of a MU that outputs motion or power is the main part of the MU and it completes the specified meta-action	In worm-worm gear transmission, the worm is the output of the element action unit in which it is located, and motion and power are transmitted to the input of the next element's action unit
Middleware	A part (part combination) that occurs between a "power input" and a "power output" and plays a major role in transmitting motion and power, and has no relative motion with the input and output parts	In worm and worm gear transmission, the coupling transmits the motion and the power output by the motor to the output of the unit - worm
Fastener	In a MU, a part that is fixed, loosen-proof, and sealed, or is used to connect two or more parts without relative motion	Such as screws, pins and end covers, spring washers, sealing rings, sealing sleeves, sealing gaskets, etc.
Supporting parts	A part in a MU that provides assembly references or supporting functions for other parts	Such as bearing, piston cylinder, sleeve, machine tool base and box

Table 1.
The definitions of MU basic elements.

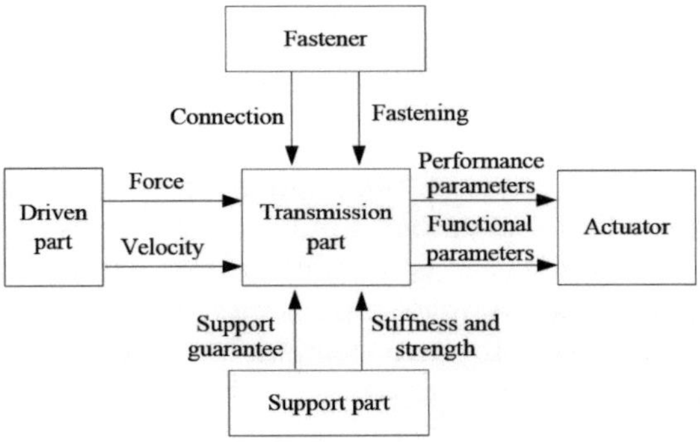

Figure 2.
The conceptual MU model.

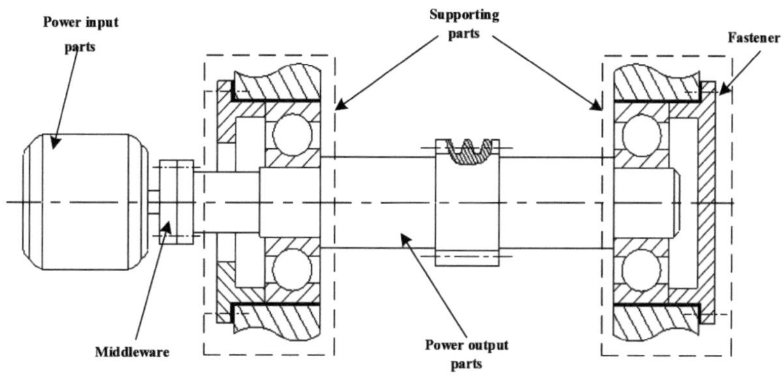

Figure 3.
Worm rotating MU.

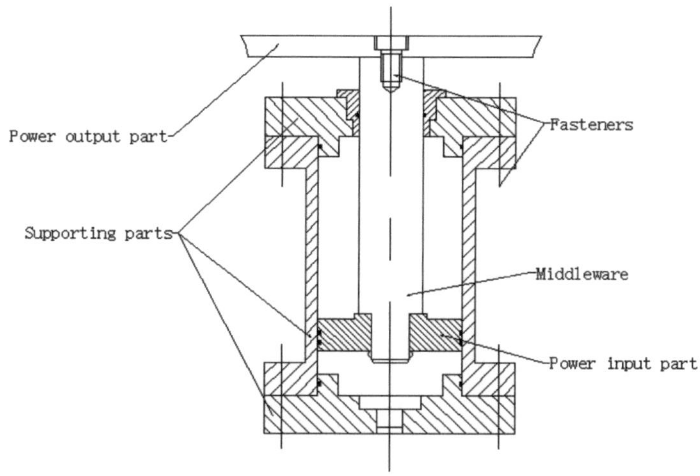

Figure 4.
Pallet moving MU.

the structural model of typical rotating units. **Figure 4** is the structural model of typical moving units.

3. The assembly model of MU

The assembly model of a MU describes the assembly process of a MU. Therefore, it is necessary to establish their standard assembly process according to the structural model of two types of MU, and draw the assembly model diagram according to the standard assembly process. **Figure 5** shows an assembly model of a MU.

3.3 Applications

3.3.1 Reliability mathematical modeling based on MU

As the smallest action unit in enabling a CNC machine tool's function, MU's reliability affects the whole system [20]. Therefore, the technology incorporated in the MU should be studied and the reliability mathematical model of the MU should be established first.

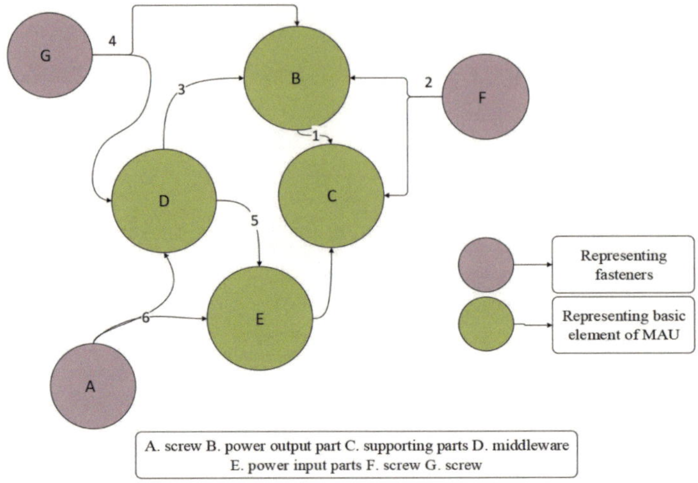

Figure 5.
The assembly model of a MU.

MU's reliability means the ability of the MU to remain functional. It can be also characterized by the degree of reliability. The reliability degree of the MU means the probability that the MU will perform its required function under given conditions for a stated time interval [21], namely, the probability that MU output characteristic parameters are within acceptable ranges in specified time periods. This is shown in Eq. (1).

$$R = P[Y_{\min} \le Y(t) \le Y_{\max}] \quad (1)$$

where, $Y(t)$ means the output quality characteristic parameters (such as precision, accuracy life, performance stability, etc.), $[Y_{\min}, Y_{\max}]$ defines the ranges of MU's output quality characteristic parameters under design requirements.

Taking the motion precision of MU, for example, and assuming that motion error values of MU follow the normal distribution, then the reliability of the MU can be described as below:

$$R = P(e_{min} \le E \le e_{max}) = P(E \le e_{max}) - P(E \le e_{min}) = \Phi\left(\frac{e_{max} - \mu}{\sigma}\right) - \Phi\left(\frac{e_{min} - \mu}{\sigma}\right)$$

In a practical situation, the CNC machine tool needs to accomplish multiple different missions by different MUs, so the system's mission reliability is actually a dynamic combination of each MU's reliability, shown in **Figure 6**.

The calculation of the machine system's mission reliability is shown in Eq. (2).

$$R^W = \sum_{i=1}^{n} \alpha_i R^{A_i} \quad (2)$$

where, R^W means the reliability of the wth mission, α_i means the weight that the ith MU relative to the wth mission, R^{A_i} means the reliability of the ith MU.

3.3.2 Reliability design for CNC machine tool based on MU

Reliability design is a basic guarantee of the CNC machine tools' reliability. As everyone knows, design of the machine tool is a difficult system engineering

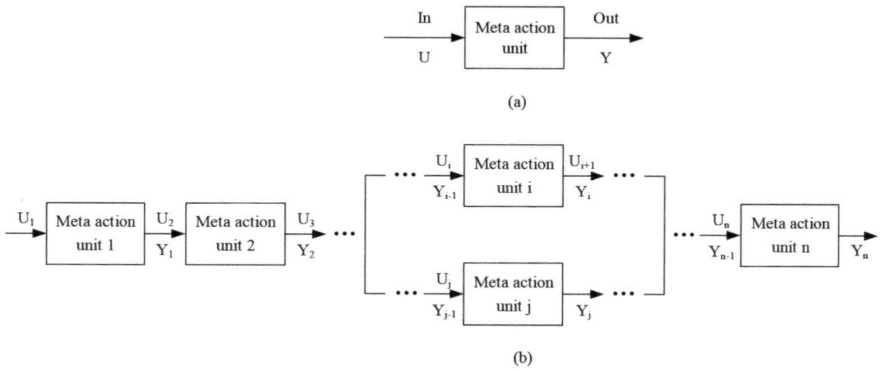

Figure 6.
Mission reliability model of MU. (a) Single meta-action unit and (b) machine system based on meta-action unit.

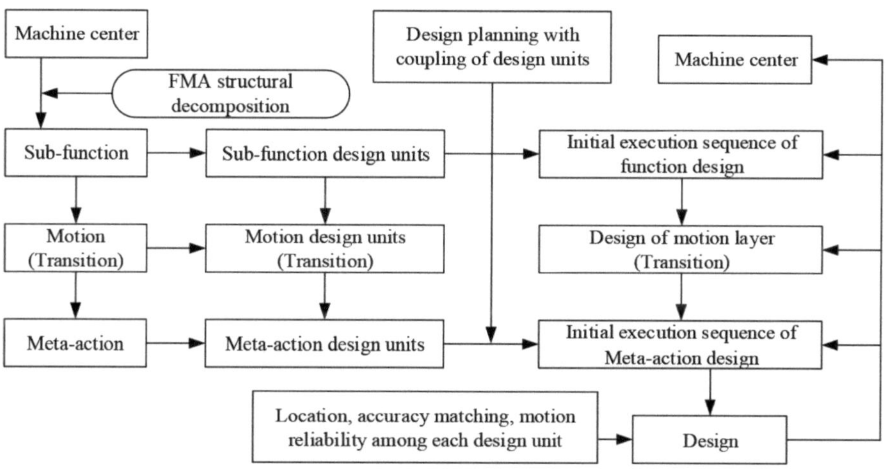

Figure 7.
Design process planning driven by FMA structural decomposition.

problem. To simplify the design, the reliability design technology by using FMA has been studied.

3.3.2.1 Design process planning for machine center driven by FMA structural decomposition

The design process of the machine center is optimized by using the FMA structural decomposition methodology [22]. There is a large amount of information coupling among each design unit; the basic planning of the design process is obtained based on the consideration of each unit's coupling, as shown in **Figure 7**, which can speed-up the design and the development of machining centers.

Firstly, the machine center is decomposed into sub-function design units, motion design units, and meta-action design units by FMA structural decomposition.

Secondly, the initial design sequence (IDS) of the function layer is obtained by considering the coupling among the design units of the sub-function layer. Next, the IDS of the meta-action layer is determined in the light of a sub-function, by taking its motion layer as a transition layer. The last step is to design the mechanical structures following an ascending order, i.e., from bottom to top (from the meta-action layer to the entire machine).

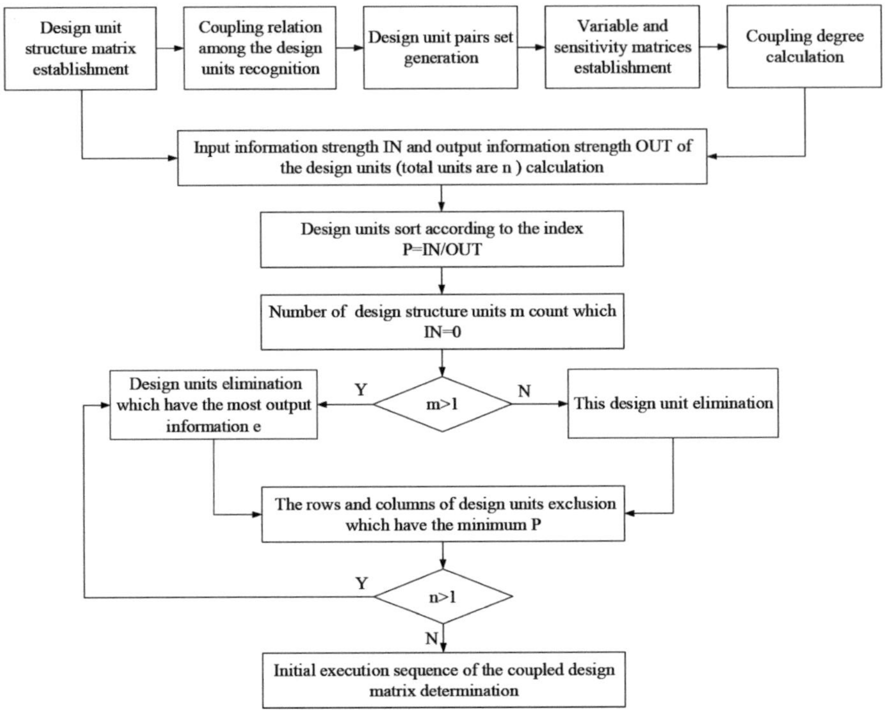

Figure 8.
Execution steps of the coupling strength calculation and coupling splitting.

After the design process planning, the coupling strengths among the design units are calculated by using variability and sensitivity indices based on the information coupling among them (i.e., the design units). Then, the splitting method is been used for the coupling design structure matrix to optimize the IDS of each design unit. **Figure 8** illustrates the procedure.

The variable stands for the degree of information change transmitted from the top design structure units to the bottom design structure units. Sensitivity means the degree of the bottom design units' output information change caused by the top design units' output information change.

3.3.2.2 Research on the evaluation of mechanical structure similarity and reliability prediction

To overcome the problem of failure data shortage because of the low yield and in order to expand the failure data, Zhang et al. [23] decomposed the CNC machine tools by FMA, and referred the failure data of similar MUs. Because of the high failure rate in the CNC machine tools, the NC rotary table is taken as an example.

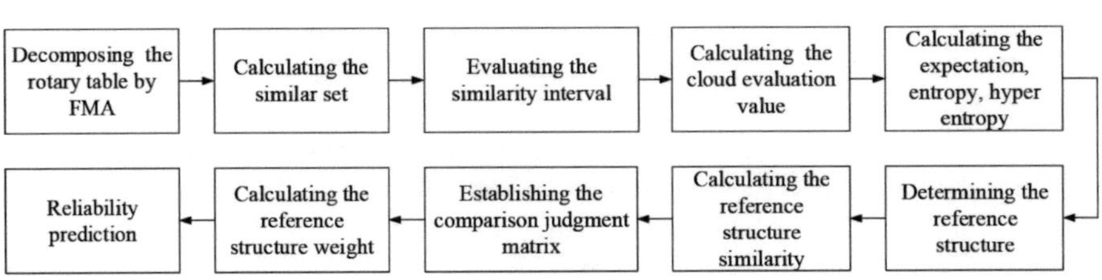

Figure 9.
Procedure of reliability prediction based on meta-action units and structure similarity.

Firstly, the NC rotary table is decomposed into the meta-action layer and possible similar unit sets of each MU are obtained. Secondly, similar MUs are determined according to the similarity of possible similar units calculated by using interval number normal cloud model. Lastly, the failure data according to the similarity among units is modified, resulting in the reliability prediction, as shown in **Figure 9**.

3.3.2.3 FTA and FMEA for meta-action unit

FTA and FMEA for meta-action unit (MU-FTA and MU-FMEA) are more suitable for the CNC machine tools that showcase the main body of mechanical structure rather than the traditional FTA and FMEA.

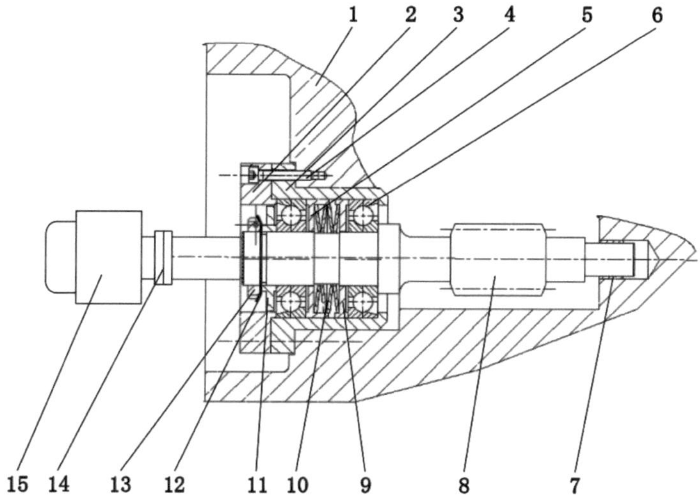

Figure 10.
Worm rotation meta-action unit: (1) slippery seat; (2) bearing cover; (3) bearing seat; (4) screw; (5) spacer sleeve; (6) bearing; (7) bushing; (8) worm; (9) spacer sleeve; (10) disk spring; (11) spacer sleeve; (12) tab washer; (13) round nut; (14) coupling; and (15) servo motor.

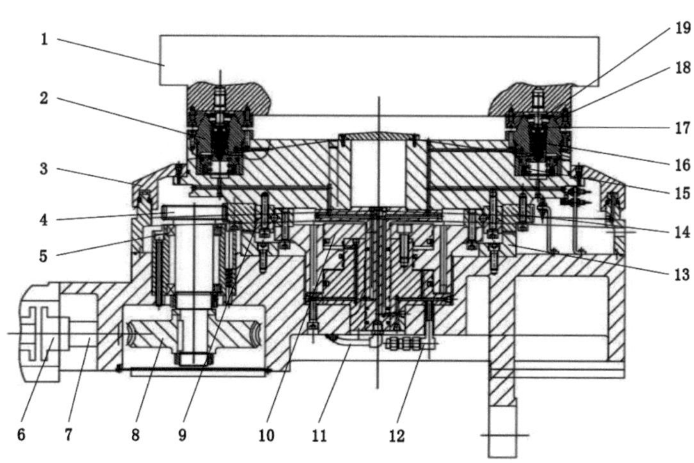

Figure 11.
End-toothed disc indexing table schematic drawing: (1) pallet; (2) male tapper; (3) sealed shell; (4) gear shaft; (5) gear shaft bearing; (6) motor; (7) worm; (8) worm gear; (9) axisymmetric body bearing; (10) lift cylinder; (11) locked cylinder oil circuit; (12) lift cylinder oil circuit; (13) lower tooth disc; (14) upper tooth disc; (15) large spring; (16) pull stud; (17) claw; (18) generating cone; and (19) positioning nail.

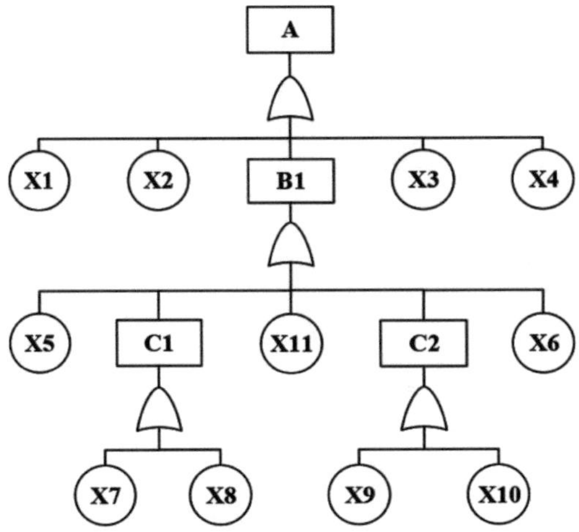

Figure 12.
FTA of worm's vibration.

Label	Event definitions	Label	Event definitions
A	Worm vibration	X5	Interference between bearing and shaft is too large
B1	Bearing vibration	X6	Fatigue failure of disc springs
C1	Bad lubrication of bearings	X7	Unreasonable grease injection
C2	Insufficient bearing preload	X8	Unclean grease
X1	Bad assembling of coupling	X9	Loosening round nut loosening
X2	Breaking liner	X10	Aging of shim gaskets
X3	Teeth bonding	X11	Bearing preload is too large
X4	Teeth pitting		

Table 2.
FTA event definition of worm vibration.

Taking the worm rotation meta-action (shown in **Figure 10**) and the end-toothed disc indexing table (shown in **Figure 11**) as examples, the MU-FTA and MU-FMEA are shown below [24]. Therefore, MU-FTA and MU-FMEA are shown in **Figure 12** and **Table 3**, and the specific contents of the fault tree are shown in **Table 2**.

3.3.3 Manufacturing technology of CNC machine tool by using FMA

Manufacturing technology is important to guarantee the reliability of CNC machine tools. Assembly, the last step of manufacture, also inadvertently affects the reliability of CNC machine tools. The research on CNC machine tools' assembly reliability by using FMA can be categorized into the following two areas:

- assembly error analysis;
- assembly reliability modeling.

3.3.3.1 Assembly error modeling technology by using FMA

The main methodologies of assembly error modeling by using FMA are assembly error transfer link graph [25] and assembly error propagation state space model [26].

Meta-action	Mode	Failure mode	Failure cause	Failure effects		Detection	Improvement measures
				Local effects	Final effects		
Worm rotation	Rotation	No action of worm rotation	Servo motor damaged	Loss worm's and gear shaft's function	Hindering the parts processing	Instruments	Maintaining or replacing the motors, couplings, bearings. Improving assembly process of bearings and couplings. Strengthening the assembly quality inspection. Inspecting the state of bearings regularly
			Coupling broken				
			Bearing jammed				
		Worm vibration	Bearing vibration	Worm and gear shaft vibration	Hindering the parts processing	Instruments	Maintaining or replacing the couplings, nuts. Improving protection and sealing structures. Changing the lubricating mode or type. Strengthening the assembly accuracy control of bearings. Improving assembly process of the bearings. Strengthening the assembly quality inspection measures. Inspecting the state of nuts and bearings regularly
			Bad assembly of bearings				
			Bearing cannot keep correct position				
			Nut loosening				
			Entry of foreign bodies				
			Bad lubrication				
			Worm root broken				
			Teeth pitting, wear or gluing				
			Wear, pitting, and gluing of upper gear disc surfaces				
			Bad assembly of large bearings				
			Bad lubrication of large bearing				
			Screw loosening				

Table 3.
MU-FMEA (partial).

1. Assembly error transfer link graph.

The assembly errors of the MU can be categorized into five aspects, namely:

- geometric position error;
- geometrical shape error;
- assembly position error;

- assembly torque (deformation) error; and
- measuring error of the parts.

The transfer and accumulation processes are shown in the assembly error transfer link graph (**Figure 18**) by using the error propagation link. The graph is a basic encapsulation unit that represents the error propagation and accumulation rules in assembly parts or between assembly parts. The function models of each part from the meta-action assembly units (MU) are presented by the circles, whereas the error propagation rules (they consist of one or several functional relations) between the function models for the parts before assembling (input) and after assembling (output) are presented by rectangles. The linkages between the function models and the rules are presented by arrows. The positive direction of the arrows is directed from the error references to the functions, shown in **Figure 13**.

For the first to the fifth error, g_{ij} means the jth function model in the part $i(i \geq 1, j \geq 0)$, d_k are the error models of the first to the fifth error, thus, the first to fifth error model $(0 \leq k \leq 5)$ of the part $i(i \geq 1, j \geq 0)$, E_{imk} means the kth error between part i and m.

There are two kinds of error propagation relation between two parts: coupling and nesting, as shown by **Figure 14**.

The complex assembly error propagation relation network (i.e., link network) is generated by the coupling and nesting of the assembly error transfer link diagram for multiple parts, shown in **Figure 15**.

At last, the link network of error propagation is transformed into the structural link matrix to predict the error propagation of MUs or the entire machine.

The link matrix is made of three aspects:

- linkage,
- error propagation model, and
- error sources.

The above are presented in **Table 4**. This methodology is used to define and describe on one hand the error source among the parts, and on the other, the error source relations during the assembly process.

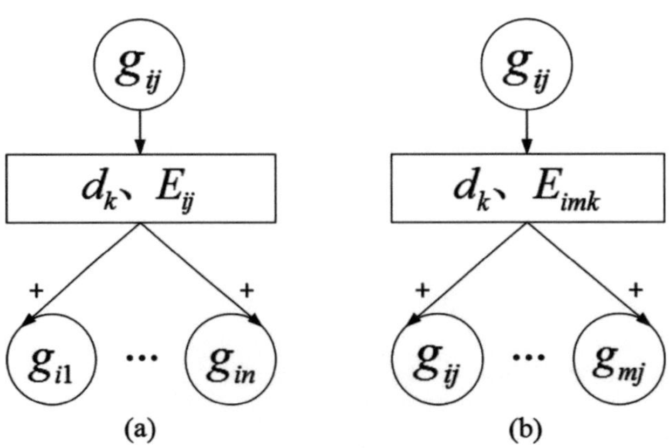

Figure 13.
Link of error propagation. (a) 1st error flow; (b) 2nd error flow.

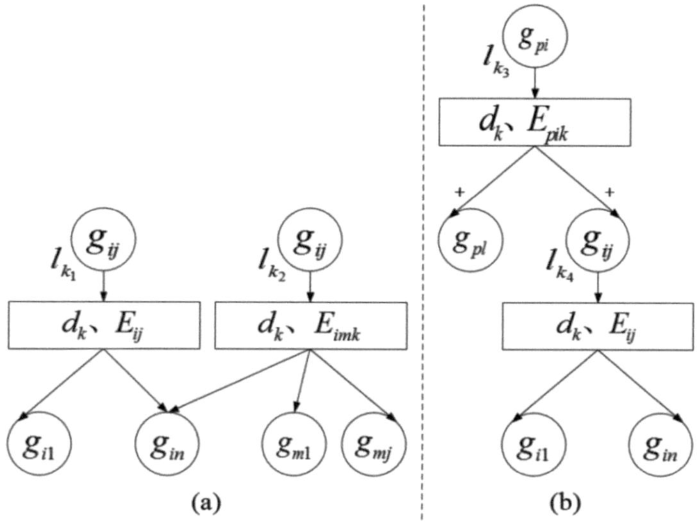

Figure 14.
Link coupling and nesting of error propagation. (a) Coupling between l_{k_1} and l_{k_2} (b) nesting between l_{k_1} and l_{k_2}.

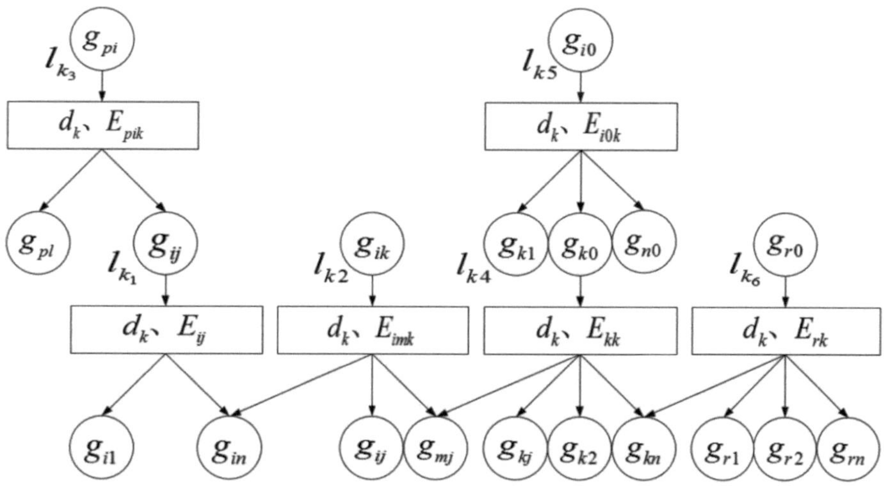

Figure 15.
Link network of the assembly error propagation (NBL).

	A					B			C							D			
	g_{pi}	g_{pl}	g_{ij}	g_{i1}	g_{in}	g_{ik}	g_{ij}	g_{mj}	g_{i0}	g_{kj}	g_{k0}	g_{n0}	g_{k1}	g_{k2}	g_{kn}	g_{r0}	g_{r1}	g_{r2}	g_{rn}
l_{k_1}				*k*	*k*														
l_{k_2}					*k*	*k*	*k*	*k*											
l_{k_3}	*k*	*k*	*k*																
l_{k_4}								*k*			*k*		*k*	*k*	*k*				
l_{k_5}									*k*	*k*	*k*	*k*							
l_{k_6}															*k*	*k*	*k*	*k*	*k*

Table 4.
Matrix of error propagation link (MBL).

The link matrix is constructed according to the two-level composite matrix architecture. In this table, the row elements represent the links, the elements in the first level column stand for the assembly parts or components, the second level column elements signify the parts contained in the components, and the center cells are identified by the error source type $k(0 \le k \le 5)$. However, if there is no error propagation or if there are no effects in assembly quality and accuracy during the error propagation, the cells should be empty.

2. State space model of assembly error propagation

The hierarchy diagram of assembly errors propagation is established by decomposing the error propagation process hierarchically based on the error propagation carriers that function assembly units, motion assembly units and MUs, is depicted in **Figure 16**.

The small displacement torsor is introduced on the basis of a hierarchy diagram, while the errors between actual geometric characteristics and ideal geometric characteristics are represented by the error vector $R = [a, b, c, \alpha, \beta, \gamma]^T$, where a, b, c and α, β, γ mean the translation errors and rotation errors along the three axes, respectively. The relative poses among assembly units are determined by their position and pose parameters, and the feature matrix is established according to the pose parameters among sub-coordinate systems, shown in Eq. (3).

$$A_k = \begin{bmatrix} 1 & -\Delta\gamma & \Delta\beta & \Delta a + x \\ \Delta\gamma & 1 & -\Delta\alpha & \Delta b + y \\ -\Delta\beta & \Delta\alpha & 1 & \Delta c + z \\ 0 & 0 & 0 & 1 \end{bmatrix} \tag{3}$$

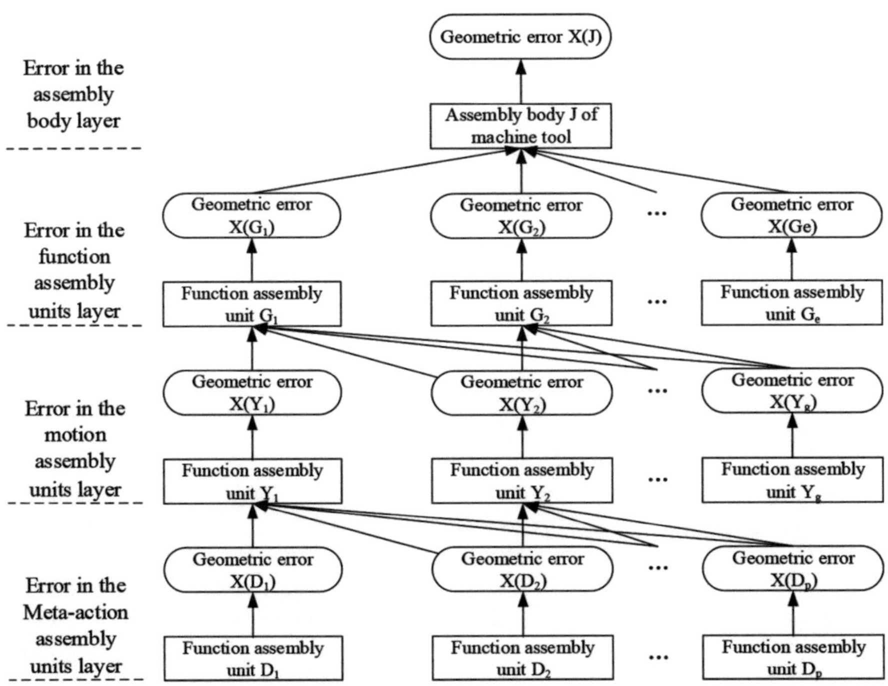

Figure 16.
Hierarchy diagram of assembly errors propagation.

Suppose the geometric characteristics of motion assembly units are affected by single factor of the MUs, thus, by sorting the MUs that affect the hth geometric error of the gth motion unit according to the assembly sequence number, shown in **Figure 17**.

According to the assembly process, after finishing the assembly of kth MU, the assembly error outputs are represented by the small displacement screw $X_{gh}(k)$ as below:

$$X_{gh}(k) = \begin{bmatrix} d_k \\ \delta_k \end{bmatrix} = \begin{bmatrix} a_k & b_k & c_k & \alpha_k & \beta_k & \gamma_k \end{bmatrix}^T$$

where, $k = 1, 2, \ldots, i$,i is the total number of the MUs that affect the hth geometric error of the gth motion unit, d_k is the translation component of geometric error, and δ_k is the rotation component of geometric error.

The errors introduced by the dynamic uncertain factors of assembly force and measurement, etc., are considered in the actual assembly process and are shown in Eq. (4).

$$\begin{cases} X_{gh}(k) = A_{gh}(k)X_{gh}(k-1) + B_{gh}(k)\mu_{gh}(k) + v_{gh}(k) \\ T_{gh}(k) = C_{gh}(k)X_{gh}(k) + \xi_{gh}(k) \end{cases} \tag{4}$$

where, $A_{gh}(k)$ is the transformation matrix of the geometric error vector among characteristic co-ordinate systems, $B_{gh}(k)$ is the error input matrix that reflects the affection of new input geometric characteristic error on assembly units, and $\mu_{gh}(k)$ is the geometric error vector introduced by the assembly of the kth MU.

The error vector consists of the errors generated by the assembly, grinding and repairing of the MUs; and $v_{gh}(k)$ is the assembly error introduced by the assembly force, $\xi_{gh}(k)$ is the measurement noise obeying the normal distribution with a mean value of 0. However, it is worth noting that there is no error input if this station

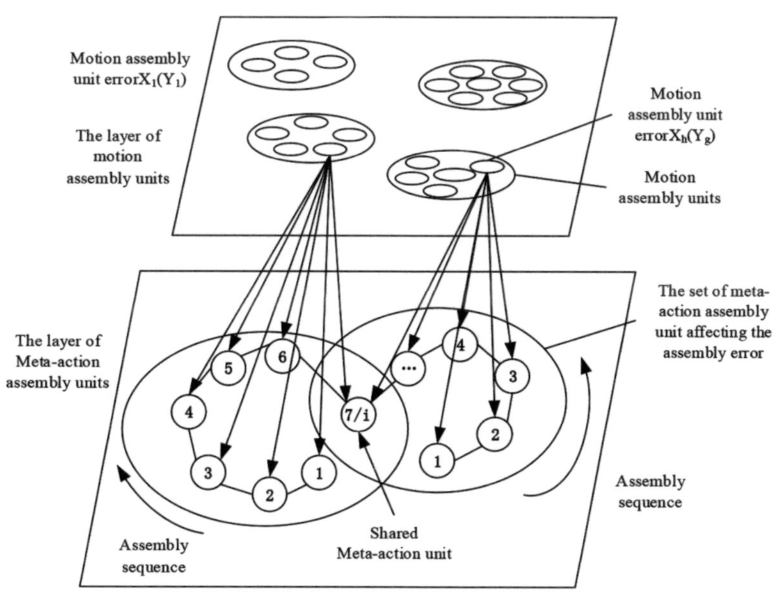

Figure 17.
Assembly process from meta-action assembly units to motion assembly units.

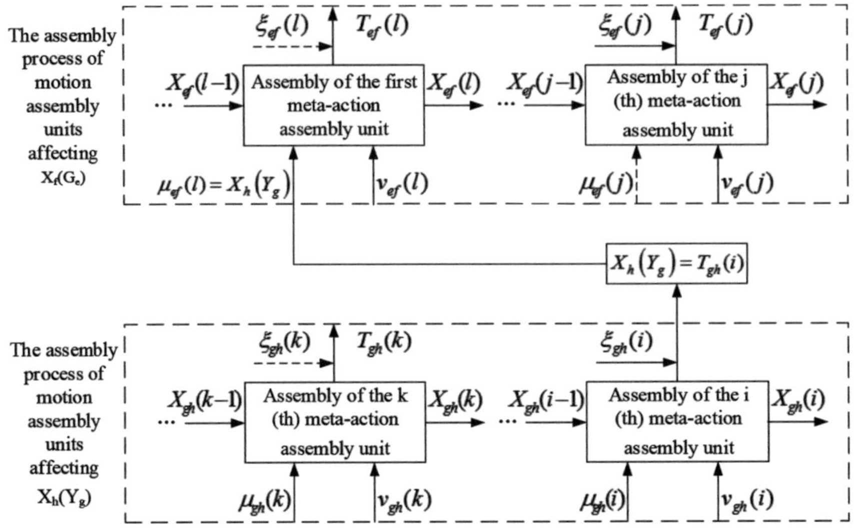

Figure 18.
The state space model of assembly error propagation.

does not need to be measured. $C_{gh}(k)$ is the output matrix and $T_{gh}(k)$ is the geometric error obtained by measuring.

The state space model of assembly error propagation is shown in **Figure 18**.

The definition of the motion assembly units' final output error is the geometric error $T_{gh}(i)$ measured after finishing the assembly of the final assembly unit i, i.e., $X_h(Y_g) = T_{gh}(i)$. Therefore, the state space models of assembly error propagating from motion assembly units to function assembly units, from function assembly units to the whole machine assembly are deduced for the same reason, shown in Eqs. (5) and (6).

$$\begin{cases} X_{ef}(k) = A_{ef}(k)X_{ef}(k-1) + B_{ef}(k)\mu_{ef}(k) + v_{ef}(k) \\ T_{ef}(k) = C_{ef}(k)X_{ef}(k) + \xi_{ef}(k) \end{cases} \tag{5}$$

$$\begin{cases} X_z(k) = A_z(k)X_z(k-1) + B_z(k)\mu_z(k) + v_z(k) \\ T_z(k) = C_z(k)X_z(k) + \xi_z(k) \end{cases} \tag{6}$$

The geometric error $X(G_e) = \{X_1(G_e), X_2(G_e), ..., X_f(G_e)\}$ of the function assembly unit e, referred as synthesis error of function assembly unit e, is obtained by introducing the error of assembly units into the state space model layer-by-layer is shown below:

$$\begin{aligned} &E(G_e) = F(X_1(G_e), X_2(G_e), ..., X_f(G_e)) = \\ &F(X_1(Y_1), X_2(Y_1)...X_1(Y_2), X_2(Y_2)...X_1(Y_n), X_2(Y_n)...) = \\ &F(X_1(D_1), X_2(D_1)...X_1(D_2), X_2(D_2)...X_1(D_n), X_2(D_n)...) \end{aligned}$$

3.3.3.2 Assembly reliability modeling based on the MUs

A large number of attempts had been made in the assembly reliability modeling of MUs, and their respective modeling methodology by the modular fault tree proposed by Li et al. [27].

The FTA is accomplished on the target product first, and decomposes the fault tree into the layer of MUs, then performs the analysis and calculation by regarding

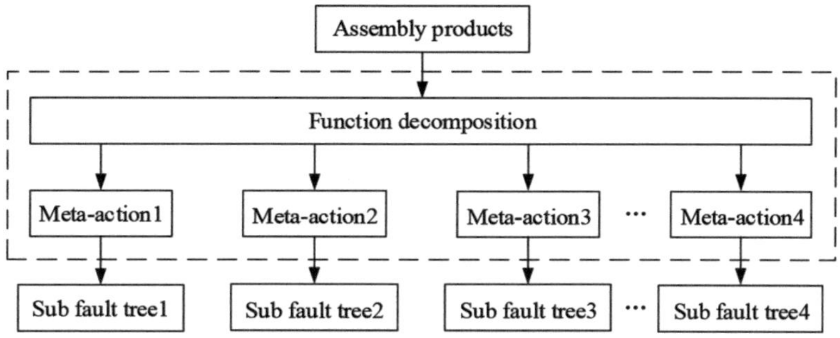

Figure 19.
Modularization fault tree model based on the function decomposition.

the meta-action fault tree after function decomposition as separate independent modules (**Figure 19**).

The function implementation is the key performance of the assembly quality; the performances of the assembly units are characterized by using the quadruples based on the modularization fault model, as $F = (S, P, T, Q)$, where:

- S symbolizes the set of assembly units' performance,
- P stands for the set of assembly units' performance attribute,
- P_m means the performance evaluation index of the assembly units, and the indices constitute the set of the assembly units' performance attribute,
- T characterizes the set of all action obtained by the function decomposition,
- T_{ij} denotes the cell of T, and $T_{i(j+1)}$ is used to represent the subordinate functional action of T_{ij} because of the inclusion relationship among the functional actions, and
- Q signifies the mapping function from the functional action to assembly performance, $Q_a : T \rightarrow P$.

On the basis of the modularization fault tree, the assembly reliability modularization fault tree is simplified by sorting basic events. Then, the fault tree is transformed into a binary decision diagram (BDD) by using ITE structural analysis methodology. Finally, transforming the meta-action sub-fault tree into a BDD, researching the assembly reliability of meta-action assembly units by combining the BDD with a mapping function, and obtaining the mapping function Q_a of meta-action assembly reliability are shown in Eq. (7).

$$Q_a : M \times F \tag{7}$$

where M means the mapping matrix of different performance attributes' weight for meta-actions and F indicates performance index evaluation results of the MUs' reliability.

3.3.4 Other reliability application based on meta-action

In addition to the application of reliability modeling, reliability design, and assembly reliability analysis for CNC machine tool based on MU, FMA methodology has also been used in failure classification [24], system motion reliability analysis [28], maintenance decision [29], to name but a few.

4. Discussion

Reliability modeling by FMA is more suitable than other decomposition methodologies because of its motion characteristics and the complicated structure of CNC machine tools. As one of the quality characteristics (i.e., precision, reliability, precision retaining ability, availability stability, and other minor characteristics), reliability is affected by other characteristics, so it is more accurate to establish the reliability model of CNC machine tools by FMA.

Reliability design is a basic assurance of a CNC machine tools' reliability. With the increasing complexity of CNC machine tools, their design became more challenging, as it was associated with poor efficiency, multiple iterations, and long design cycles. The entire machine was decomposed into MUs to ensure accomplishing the function by means of simple rotations and movements. As such, the design of the entire machine is turned into the design of MUs, and the design process of the CNC machines is hence simplified.

Otherwise, practice has shown that the traditional similar product method, which seeks for similar structure at the whole machine level, in conjunction with FMA can expand the failure data more accurately, thereby reaching more precise conclusions. As such, an FMA decomposition can simplify the CNC machine tools by making the analysis more efficient and avoiding duplicate results. Otherwise, since traditional FTA and FMEA are carried out on the basis of the parts, MU-FTA and MU-FMEA can reduce the number of agreed levels and reduce the workload.

As far as the assembly technology is concerned, it would cause data explosion and increase the difficulty of analysis from the research of assembly-specific technology based on the parts. In the entire machine layer, the assembly technology research will be affected by the coupling relationship between the parts, which increases the disassembly difficulty of the assembly process. To reduce the difficulty of reliability assembly work, the common approach is to simplify the products by decomposition. FMA decomposition is more suitable for the assembly reliability study than other decomposition methodology, because of the quality characteristic similarity between the complete machine and the units.

5. Conclusions and further work

Performed research showed that meta-action methodology is adaptive and scientifically correct for the reliability analysis of the product. In this chapter, the meta-action methodology is introduced. To obtain the meta-action, the FMA decomposition process and its rules were presented. The meta-action and their corresponding units were defined and the constituent parts of the meta-action unit were shown. Some applications were introduced, such as reliability model, allocation, assessing, and fault diagnosing.

Meta-action methodology, as a new kind of structural decomposition theory, is more suitable for quality and reliability analysis of mechanical systems than traditional methods. It is an important tool to accomplish the reliability related work for CNC machine tools and even electromechanical products.

It is the authors' view that it will be more widely used in the future based on its constant deep study. In view of the above, research on reliability based on meta-action should be further facilitated and performed. A systematic research method of reliability based on meta-action method can be built, which would promote the reliability level of CNC machine tool holistically. This basically includes the following three aspects:

- reliability design technology from bottom to top by regarding the meta-actions as the smallest units, since the meta-actions are decomposed from functions;
- fault mode classification by meta-action, because the fault modes of meta-action units are relatively fixed and have certain regularity; and
- fault mechanism study by meta-action, as the FMA has the function of simplifying the CNC machine tools.

Author details

Yan Ran*, Wei Zhang, Zongyi Mu and Genbao Zhang
Chongqing University, China

*Address all correspondence to: ranyan@cqu.edu.cn

References

[1] Zhang Y, Feng R, Shen G, et al. Weight calculation for availability demand index of CNC machine tools based on market competition and self-correlation. Computer Integrated Manufacturing Systems. 2016;**22**(07): 1687-1694

[2] Whiteley M, Dunnett S, Jackson L. Failure mode and effect analysis, and fault tree analysis of polymer electrolyte membrane fuel cells. The International Journal of Hydrogen Energy. 2016;**41**: 1187-1202. DOI: 10.1016/j.ijhydene.2015.11.007

[3] Zhang H, Liu B. Integrated analysis of software FMEA and FTA. International Conference on Information Technology and Computer Science. 2009;**2**:184-187. DOI: 10.1109/ITCS.2009.254

[4] Ng WC, Teh SY, Low HC, et al. The integration of FMEA with other problem solving tools: A review of enhancement opportunities. Journal of Physics Conference Series. 2017;**890**: 012139. DOI: 10.1088/1742-6596/890/1/012139

[5] Zhu XC, Chen F, Li XB, et al. Key subsystem identification of the CNC machine tools. Applied Mechanics and Materials. 2013;**329**:157-162. DOI: 10.4028/www.scientific.net/AMM.329.157

[6] Karyagina M, Wong W, Vlacic L. Reliability aspect of CNC machines-are we ready for integration. In: IEEE Symposium on Emerging Technologies and Factory Automation, Etfa. IEEE; 1995. DOI: 10.1109/ETFA.1995.496736

[7] Su C, Xu YQ. Research on theory and methods of dynamic reliability modeling for complex electromechanical products. Manufacture Information Engineering of China. 2005;**35**(9):24-32

[8] Zhang GB, Wang GQ, et al. Research on reliability allocation technology of CNC machine tools based on task. China Mechanical Engineering. 2010;**21**(19): 2269-2273

[9] Xin ZJ, Xu YS, et al. FEM-Based static and dynamic design of numerical control gear machining tool column. Journal of North University of China (Natural Science Edition). 2006;**27**(06): 483-486

[10] Zhang GB, Ge HY, et al. Assembly process modeling and prediction method of reliability-driven. Computer Integrated Manufacturing Systems. 2010;**18**(2):349-355

[11] Zhang GB et al. Research on reliability-driven assembly process modeling method. Journal of Agricultural Machinery. 2011;**42**(10): 192-196

[12] Jing O, Jianghong Z, Hao T. Research and application on industrial product innovative design base on Scenario-FBS model. International Symposium on Computational Intelligence and Design. 2011

[13] Li B. Research on Product Configuration and Assembly Line Optimal Scheduling for Mass Customization. China: Mechanical Engineering, Huazhong University of Science and Technology; 2007

[14] Zhang W, Zhang GB, Ran Y, Shao YM. The full-state reliability model and evaluation technology of mechatronic product based on meta-action unit. Advances in Mechanical Engineering. 2018;**10**(4):1-11

[15] Jia ZX, Zhang HB, Yi AM. Expanding reliability data of CNC machine tools by using neural networks. Journal of Jilin University (Engineering

Edition). 2011;**41**(2):403-407. DOI: 10.3901/JME.2010.02.145

[16] Zhang GB, Zhang L, Ran Y. Reliability and failure analysis of CNC machine based on element action. Applied Mechanics and Materials. 2014; **494-495**:354-357

[17] Ao MY, Zhao Y. The research status and development trend of fault diagnosis system for CNC machine. Applied Mechanics and Materials. 2012; **229-231**(3):2229-2232

[18] Lu X, Gao S, Han P. Failure mode effects and criticality analysis (FMECA) of circular tool magazine and ATC. Journal of Failure Analysis and Prevention. 2013;**13**(2):207-216. DOI: 10.1007/s11668-013-9654-9

[19] Zhang HB, Jia ZX, Yun AM. Fuzzy decision method for reliability allocation of NC machine tools. Manufacturing Technology and Machine Tools. 2009;**2**: 60-63

[20] Ran Y. Research on meta-action unit modelling and key QCs predictive control technology of electromechanical products. Chong Qing University. 2016: 53-55

[21] Birolini A. Reliability Engineering: Theory and Practice. 7th ed. Springer; 2014. 2 p. DOI: 10.1007/978-3-642-39535-2

[22] Zhu GY, Liu Y, Zhang GB, Xia CJ. Machining center design process planning driven by FMA structural decomposition. Mechanical Science and Technology for Aerospace Engineering. 2017;**36**(8):1167-1174

[23] Zhang GB, Xu FW, Ran Y, Zhang XG. Research on similarity evaluation and reliability prediction of mechanical structure. Chinese Journal of Engineering Design. 2017;**24**(3):264-272

[24] Yao MS. Research on reliability analysis technology of typical meta-action units of NC machine tools. Chong Qing University. 2018;**25-34**:70-71

[25] Li DY, Li MQ, Zhang GB, Wang Y, Ran Y. Mechanism analysis of deviation sourcing and propagation for meta-action assembly unit. Journal of Mechanical Engineering. 2015;**51**(17): 146-155

[26] Sun YY, Liu Y, Ran Y, Zhou QF. Assembly precision prediction method of numerical control machine tools based on meta-action. Mechanical Science and Technology for Aerospace Engineering. 2017;**36**(11):1734-1739

[27] Li DY. Research on quality modeling and diagnosis technology for the assembly process of CNC machine tool. Chong Qing University. 2014:123-132

[28] Zhang GB, Zhang H, Fan XJ, Tu L. Function decomposition and reliability analysis of CNC machine using function-motion-action. Mechanical Science and Technology for Aerospace Engineering. 2012;**31**(4):528-533

[29] Zhang GB, Yang XY, Li DY, Li L. Failure maintenance decision of meta-action assembly unit. Mechanical Science and Technology for Aerospace Engineering. 2016;**35**(5):722-728

Chapter 4

Reliability Analysis Based on Surrogate Modeling Methods

Qian Wang

Abstract

Various surrogate modeling methods have been developed to generate approximate functions of expensive numerical simulations. They can be used in reliability analysis when integrated with a numerical reliability analysis method such as a first-order or second-order reliability analysis method (FORM/SORM), or Monte Carlo simulations (MCS). In this chapter, a few surrogate modeling methods are briefly reviewed. A reliability analysis approach using surrogate models based on radial basis functions (RBFs) and successive RBFs is presented. The RBF surrogate modeling method is a special type of interpolation method, as the model passes through all available sample points. Augmented RBFs are adopted to create approximate models of a limit state/performance function, before the failure probability can be computed using MCS. To improve model efficiency, a successive RBF (SRBF) surrogate modeling method is investigated. Several mathematical and practical engineering examples are solved. The failure probabilities computed using the SRBF surrogate modeling method are fairly accurate, when a reasonable sample size is used to create the surrogate models. The method based on augmented RBF surrogate models is useful for probabilistic analysis of practical problems, such as civil and mechanical engineering applications.

Keywords: reliability analysis, surrogate models, successive radial basis function (SRBF), failure probability, Monte Carlo simulations (MCS)

1. Introduction

The probabilistic analysis of practical engineering problems has been a traditional research field [1–3]. The first category of engineering reliability analysis methods are the most probable point (MPP) methods [4–7]. In this category of methods, a design point, or the so-called most probable point in the design space is sought. The limit state function is often transformed into a standard Gaussian space and approximated using Taylor series expansions. Depending on the order of approximation used, FORM/SORM are available [4–7]. These methods require the derivatives of system responses, i.e., sensitivity analysis. For complex engineering systems that require expensive response simulations such as nonlinear explicit finite element (FE) analysis, the integration of the MPP-based methods and a commercial FE code is not straightforward. An alternative category of methods are the direct sampling-based methods, including MCS and some other simulation methods [8–12]. These methods can be integrated fairly easily with an existing simulation program because they do not require the derivation or calculation of gradient information. However the direct application of MCS can be computationally prohibitive in complex engineering problems that require expensive response simulations.

To reduce the complexity of implementation and improve the computational efficiency, various approximate modeling techniques have been applied to the reliability analysis of practical engineering systems [13, 14]. These approximate models are referred to as surrogate models. There are abundant literature that presented surrogate models and their applications to numerical optimization and reliability-based design optimization. However, the focus of this chapter and the review of literature here is primarily on the applications of surrogate models to engineering reliability analysis. In surrogate modeling methods, the analysis software is replaced by approximate surrogate models, which have explicit functions and are very efficient to evaluate. FORM/SORM or a sampling method can then be applied using the explicit surrogate model instead of the original implicit numerical model. In all the surrogate models developed, the most basic and popular surrogate model is the conventional polynomial-based response surface method (RSM). The RSM has been shown to be useful for different engineering reliability analyses and applications [15–25]. The entire response space is approximated using a single quadratic polynomial function in a global RSM model. To improve model accuracy for reliability analysis using a global RSM model, different techniques were proposed such as efficient sampling methods [26, 27] and inclusion of higher order effects [28, 29]. When combined with gradient-based search methods, it is more efficient to use RSM in an iterative manner or a local window of the response space [30]. Local RSM methods such as the moving least square technique were developed to handle highly nonlinear limit state functions [31]. Other commonly used surrogate modeling methods have also been developed over the years, such as artificial neural networks (ANN) [32–37], Kriging [38–46], high-dimensional or factorized high-dimensional model representation [47–51], support vector machine [52–57], radial basis functions (RBFs) [58], and even ensemble of surrogates [59–62].

An RBF surrogate model is a multidimensional interpolation approach using available scattered data. Due to their characteristics in global approximation, RBFs could create accurate surrogate models of various responses [63, 64]. An RBF model provides exact fit at the sample points. In the studies by Fang and coauthors [65, 66], various basis functions were investigated including Gaussian, multiquadric, inverse multiquadric, and spline functions. Some compactly supported (CS) basis functions developed by Wu [67] were also studied. Mathematical functions and practical engineering responses were tested and their surrogate models were created using different basis functions. Augmented compactly supported functions worked well and were found to create more accurate surrogate models than non-augmented models.

2. Aims and objectives

It can be seen from literature review that accurate and efficient surrogate models are useful tools when integrated with expensive response simulations for practical reliability analysis and design problems. The objective of this research is to study efficient and accurate RBF models, such as adaptive or successive RBF models based on the augmented basis functions, and their application in engineering reliability analysis. Note that the accuracy of RBF surrogate models depends on the sample size used. If the sample size is too small, the model may not be accurate. On the other hand, a large number of sample points will improve the model accuracy, but some sample points and associated response simulations may not be necessary.

Since the most appropriate sample size is not known before the creation of the surrogate models, it remains a challenge to determine the appropriate sample size to use. One viable approach is to create and test a few different sample sizes, and the best sample size for the problem can be determined. To improve this process, the concept of SRBF surrogate models is developed and it is intended to automate this process and find the proper sample size iteratively and automatically for the augmented RBF surrogate models that can be used for reliability analysis of practical engineering systems.

This chapter presents an engineering reliability analysis method based on a SRBF surrogate modeling technique. In each iteration of the new method, augmented RBFs can be used to generate surrogate models of a limit state function. Three accurate augmented RBFs surrogate models, which were identified from a previous study, are adopted. The failure probability can be calculated using the SRBF surrogate models combined with MCS. Section 3 describes the general concept of engineering reliability analysis. Section 4 briefly reviews some surrogate modeling methods, and explains the augmented SRBF surrogate modeling technique. Sections 5 and 6 presents the MCS method and the overall reliability analysis procedures. In Section 7, the proposed approach is applied to the probability analysis of several mathematical and practical engineering problems. The failure probabilities are compared with those computed based on the direct implementation of MCS without surrogate models. The numerical accuracy and efficiency of the proposed approach using MCS and SRBF surrogate models is studied.

3. Engineering reliability analysis

A time-invariant reliability analysis of an engineering problem is to compute the failure probability, P_F, using the following integral [1–3]:

$$P_F \equiv P(g(\mathrm{x}) \le 0) = \int_{g(\mathrm{x}) \le 0} p_X(\mathrm{x}) d\mathrm{x} \tag{1}$$

where x is an s-dimensional real-valued vector of random variables, $g(\mathrm{x})$ is the limit state function, and $p_X(\mathrm{x})$ is the joint probability density function. Eq. (1) is difficult to obtain for practical engineering applications, since $p_X(\mathrm{x})$ is unknown and $g(\mathrm{x})$ is usually an implicit and nonlinearity function. A detailed response analysis model, such as the FE analysis of the engineering system is often required to evaluate function values of $g(\mathrm{x})$.

4. Surrogate modeling methods

4.1 Design of experiments

An implicit function $g(\mathrm{x})$ is considered, where $\mathrm{x} = [x_1 \ \cdots \ x_s]^{\mathrm{T}}$ is an input variable vector and s is the number of input variables. Before a surrogate model of function $g(\mathrm{x})$ can be created, some sample points shall be generated using design of experiments (DOE). Some routinely used DOE approaches include factorial design, Latin hypercube sampling (LHS) [68], central composite design, and Taguchi orthogonal array design [69]. Assume x_i is the input variable vector at the ith (i = 1,...n) sample point, the limit state function $g(\mathrm{x})$ needs to be evaluated at all the sample points to obtain the function values, i.e., $g = [g_1 \ \cdots \ g_n]^{\mathrm{T}} = [g(\mathrm{x}_1) \ \cdots \ g(\mathrm{x}_n)]^{\mathrm{T}}$.

4.2 Response surface method using quadratic polynomials

Using linear or quadratic polynomials, a response surface model can be developed. The most commonly used quadratic polynomial response surface model is expressed as [63]:

$$\widetilde{g}(\mathrm{x}) = \beta_0 + \sum_{i=1}^{s} \beta_i x_i + \sum_{i=1}^{s} \beta_{ii} x_i^2 + \sum_{i=1}^{s-1} \sum_{j=i+1}^{s} \beta_{ij} x_i x_j \tag{2}$$

where the β's are the unknown coefficients. Using the function values at n sample points, a total of n linear equations can be written in a matrix form, as:

$$g = X\widetilde{\beta} \tag{3}$$

where $\widetilde{\beta}$ $(k \times 1)$ is the least-square estimation of the unknown coefficients in Eq. (2), and X $(n \times k)$ is a matrix of input variables at sample points. Apply the least squares method to solve for $\widetilde{\beta}$, as:

$$\widetilde{\beta} = \left(X^T X\right)^{-1} \left(X^T g\right) \tag{4}$$

4.3 Least squares support vector machine

The support vector machine (SVM) uses a nonlinear mapping technique and solves for a nonlinear input-output relationship. For n sample points, a commonly used least squares SVM model is given as [52, 53]:

$$\widetilde{g}(\mathrm{x}) = \sum_{i=1}^{n} \alpha_i K(\mathrm{x}, \mathrm{x}_i) + b \tag{5}$$

where α_i $(i = 1,\dots n)$ are Lagrange multipliers, b is the scalar threshold, and $K(\mathrm{x}, \mathrm{x}_i)$ is a kernel function. Available kernel functions include polynomial, radial, and sigmoid kernels [53]. A system of $(n + 1)$ equations can be written as:

$$\begin{pmatrix} 0 & 1^T \\ 1 & \Omega + \gamma^{-1} I \end{pmatrix} \begin{pmatrix} b \\ \alpha \end{pmatrix} = \begin{pmatrix} 0 \\ g \end{pmatrix} \tag{6}$$

where γ is a tolerance error, $1 = \begin{bmatrix} 1 & \cdots & 1 \end{bmatrix}^{\mathrm{T}}$, $\alpha = \begin{bmatrix} \alpha_1 & \cdots & \alpha_n \end{bmatrix}^{\mathrm{T}}$, and Ω $(n \times n)$ is a matrix of kernels based on the sample points. α and b can be calculated from:

$$\begin{pmatrix} b \\ \alpha \end{pmatrix} = \begin{pmatrix} 0 & 1^T \\ 1 & \Omega + \gamma^{-1} I \end{pmatrix}^{-1} \begin{pmatrix} 0 \\ g \end{pmatrix} \tag{7}$$

4.4 Kriging

The Kriging model is an interpolation technique that combines two parts, i.e., a linear regression part and a stochastic error, as [38, 39]:

$$\widetilde{g}(\mathrm{x}) = B^T(\mathrm{x})\beta + z(\mathrm{x}) = \sum_{i=1}^{p} B_i(\mathrm{x})\beta_i + z(\mathrm{x}) \tag{8}$$

where $B(\mathrm{x}) = \begin{bmatrix} B_1(\mathrm{x}) & \cdots & B_p(\mathrm{x}) \end{bmatrix}^{\mathrm{T}}$ are the p basis functions, and $\beta = \begin{bmatrix} \beta_1 & \cdots & \beta_p \end{bmatrix}^{\mathrm{T}}$ are the corresponding regression coefficients. The first part of Eq. (8) approximates the global trend of the original function, in which β can be estimated using the least squares method. The second part, $z(\mathrm{x})$, represents a stochastic process with zero mean and covariance

$$Cov\left[z(\mathrm{x}_i), z(\mathrm{x}_j)\right] = \sigma^2 \mathrm{R}\left[R\left(\mathrm{x}_i, \mathrm{x}_j\right)\right] \tag{9}$$

where σ^2 is the process variance, and R is a correlation matrix. If Gaussian function is used as the correlation function, $R\left(\mathrm{x}_i, \mathrm{x}_j\right)$ is written as:

$$R\left(\mathrm{x}_i, \mathrm{x}_j\right) = \exp\left[-\sum_{k=1}^{s} \theta_k \left|\mathrm{x}_i^k - \mathrm{x}_j^k\right|^2\right]^* \tag{10}$$

where x_i^k and x_i^k are the kth (k = 1,... s) component of sample points x_i and x_j, respectively, and θ_k are unknown correlation parameters to fit the model.

4.5 Augmented radial basis functions

Developed for fitting topographic contours, an RBF surrogate model $\widetilde{g}(\mathrm{x})$ is written as:

$$\widetilde{g}(\mathrm{x}) = \sum_{i=1}^{n} \lambda_i \phi(\|\mathrm{x} - \mathrm{x}_i\|) \tag{11}$$

where ϕ is the basis function, $\|\mathrm{x} - \mathrm{x}_i\|$ is the Euclidean norm, and λ_i is the unknown weighted coefficient that need to be determined. **Table 1** lists commonly used RBFs.

Using the n available sample points and function values, a total of n equations can be written, as:

$$g_1 = \widetilde{g}(\mathrm{x}_1) = \sum_{i=1}^{n} \lambda_i \phi(\|\mathrm{x}_1 - \mathrm{x}_i\|) \tag{12}$$

...

$$g_n = \widetilde{g}(\mathrm{x}_n) = \sum_{i=1}^{n} \lambda_i \phi(\|\mathrm{x}_n - \mathrm{x}_i\|) \tag{13}$$

Write all the n equations in a matrix form, as:

$$g = A\lambda \tag{14}$$

Function name	Radial basis function
Linear function	$\phi(r) = r$
Cubic function	$\phi(r) = r^3$
Gaussian function	$\phi(r) = exp\left(-cr^2\right);\ 0 < c \leq 1$
Multiquadric function	$\phi(r) = \sqrt{r^2 + c^2};\ 0 < c \leq 1$
CS function $\phi_{2,0}$	$\phi_{2,0}(z) = (1-z)^5(1 + 5z + 9z^2 + 5z^3 + z^4);\ z = r/r_0$
CS function $\phi_{2,1}$	$\phi_{2,1}(z) = (1-z)^4(4 + 16z + 12z^2 + 3z^3)$
CS function $\phi_{3,0}$	$\phi_{3,0}(z) = (1-z)^7(5 + 35z + 101z^2 + 147z^3 + 101z^4 + 35z^5 + 5z^6)$
CS function $\phi_{3,1}$	$\phi_{3,1}(z) = (1-z)^6(6 + 36z + 82z^2 + 72z^3 + 30z^4 + 5z^5)$

Table 1.
Some commonly used RBFs [65].

where $\lambda = [\lambda_1 \quad \cdots \quad \lambda_n]^{\mathrm{T}}$, and A is given as:

$$A = \begin{bmatrix} \phi(\|\mathrm{x}_1 - \mathrm{x}_1\|) & \cdots & \phi(\|\mathrm{x}_1 - \mathrm{x}_n\|) \\ \vdots & \ddots & \vdots \\ \phi(\|\mathrm{x}_n - \mathrm{x}_1\|) & \cdots & \phi(\|\mathrm{x}_n - \mathrm{x}_n\|) \end{bmatrix}_{n\times n} \tag{15}$$

Solve the linear system of Eq. (14) to calculate coefficients λ, as:

$$\lambda = A^{-1}g \tag{16}$$

Since highly nonlinear basis functions are used, the RBF surrogate models in Eq. (11) can approximate nonlinear responses very well. However, they were found to have more errors for linear responses [58]. In order to overcome this drawback, the RBF model in Eq. (11) can be augmented by polynomial functions, as:

$$\tilde{g}(\mathrm{x}) = \sum_{i=1}^{n}\lambda_i\phi(\|\mathrm{x} - \mathrm{x}_i\|) + \sum_{j=1}^{p}c_jf_j(\mathrm{x}) \tag{17}$$

where the second part represents p terms of polynomial functions, and c_j ($j = 1,\ldots p$) are the unknown coefficients to be determined. There are more unknowns than available equations; therefore the following orthogonality condition is required to solve for all unknowns, as:

$$\sum_{i=1}^{n}\lambda_if_j(\mathrm{x}_i) = 0, \ \text{for}\, j = 1, \ldots p \tag{18}$$

Eqs. (17) and (18) consist of $(n + p)$ equations in total, and they can be rewritten, as:

$$\begin{pmatrix} A & F \\ F^T & 0 \end{pmatrix}\begin{pmatrix} \lambda \\ c \end{pmatrix} = \begin{pmatrix} g \\ 0 \end{pmatrix} \tag{19}$$

where $c = [c_1 \quad \cdots \quad c_p]^{\mathrm{T}}$, and F is given as:

$$F = \begin{bmatrix} f_1(\mathrm{x}_1) & \cdots & f_p(\mathrm{x}_1) \\ \vdots & \ddots & \vdots \\ f_1(\mathrm{x}_n) & \cdots & f_p(\mathrm{x}_n) \end{bmatrix}_{n\times p} \tag{20}$$

Solve the linear system of Eq. (19) to get λ and c, as:

$$\begin{pmatrix} \lambda \\ c \end{pmatrix} = \begin{pmatrix} A & F \\ F^T & 0 \end{pmatrix}^{-1}\begin{pmatrix} g \\ 0 \end{pmatrix} \tag{21}$$

For augmented RBFs, either linear or quadratic polynomial functions can be used. In this study, only linear polynomial functions were added to Eq. (17). For the rest of the paper, a suffix "-LP" is used to represent linear polynomials added to RBFs. The following RBF models were studied:

SRBF-MQ-LP: sequential multiquadric function with linear polynomials.

SRBF-CS20-LP: sequential compactly supported function $\phi_{2,0}$ with linear polynomials.

SRBF-CS30-LP: sequential compactly supported function $\phi_{3,0}$ with linear polynomials.

5. Estimation of failure probability

Eqs. (11) and (17) are the RBF and augmented RBF surrogate model functions of $g(\mathrm{x})$. The surrogate models have explicit expressions; therefore their function values can be efficiently calculated in each iteration of the SRBF approach. Based on the surrogate model $\tilde{g}(\mathrm{x})$, the failure probability P_F can be computed using a sampling method, such as MCS, as:

$$P_F \equiv P(g(\mathrm{x}) \leq 0) = \frac{1}{N}\sum_{i=1}^{N} \Gamma\left[\tilde{g}\left(\mathrm{x}^i\right) \leq 0\right] \tag{22}$$

where N is the total number of MCS samples, x^i is the ith realization of x, and Γ is a deciding function, as:

$$\Gamma = \begin{cases} 1 \; if \; \tilde{g}\left(\mathrm{x}^i\right) \leq 0 \\ 0 \; if \; \tilde{g}\left(\mathrm{x}^i\right) > 0 \end{cases} \tag{23}$$

The reliability index β can be further determined, as [49]:

$$\beta = -\Phi^{-1}(P_F) \tag{24}$$

where Φ is the standard normal cumulative distribution function.

6. Reliability analysis based on successive RBF models

Figure 1 shows a flowchart of reliability analysis using SRBF-based surrogate modeling technique and MCS. Once the explicit augmented RBF surrogate model is generated in one iteration of the proposed method, MCS is applied to efficiently estimate the failure probability for any sample size. If the convergence criterion is not satisfied in the current iteration, more sample points will be added and another iteration starts. As the sample size increases, the SRBF surrogate models in general become more accurate, a reduction was observed in the failure probability estimation errors. However this results in more function evaluations. Since the number of response simulations is determined by the sample size used to create a surrogate model, the majority of the computational cost is from the response simulations. The detailed procedure is as follows:

1. Determine initial and additional sample sizes, n and m, and convergence criterion. In this study, the initial sample size n is suggested be 5–10 times of the number of random variables s. The additional sample size m in each subsequent iteration can be typically taken as one third to one half of the initial sample size, n.

2. Generate the initial sample set with n sample points; set the iteration number $k = 1$. A commonly used LHS was applied to generate samples for RBF surrogate models.

3. Evaluate limit state function $g(\mathrm{x})$ for the initial sample set n generated in Step 2. Numerical analyses such as FE analyses may be required for practical problems.

4. Update sample set n to include all sample points, $n = n + m$. For the first iteration ($k = 1$), $m = 0$, and no additional sample points are added.

5. Construct augmented RBF surrogate models $\tilde{g}(\mathrm{x})$ of function $g(\mathrm{x})$ based on Eq. (17) using all available sample points.

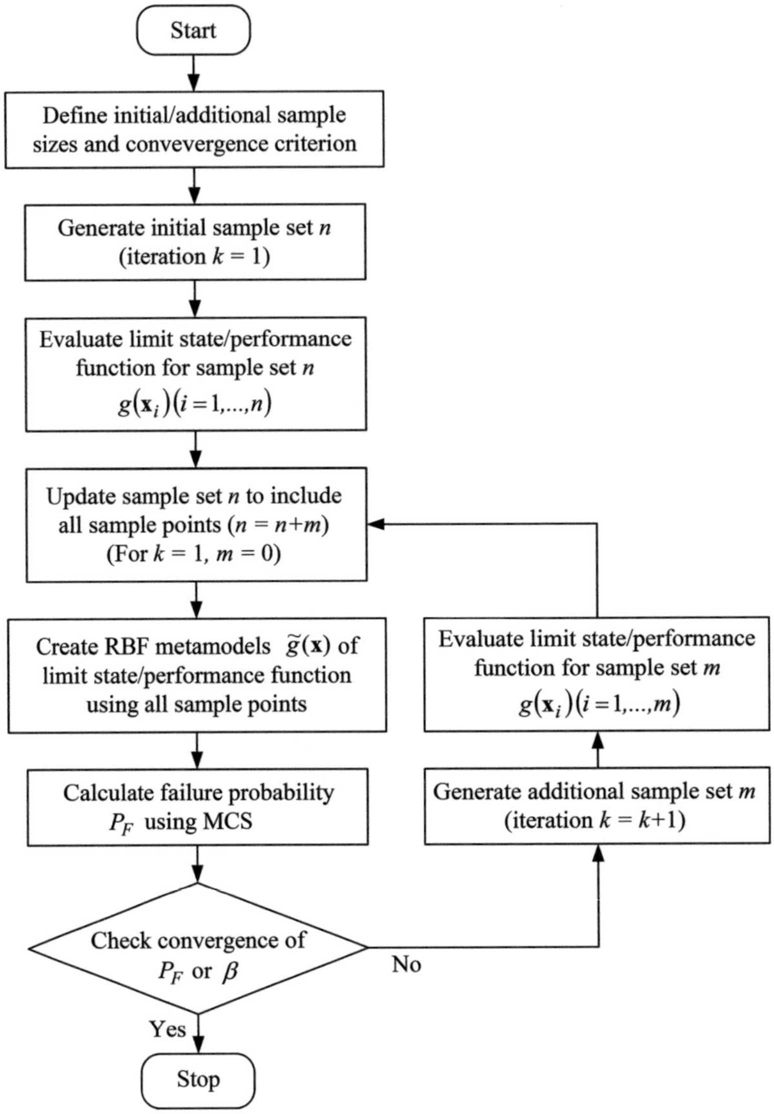

Figure 1.
Flowchart of reliability analysis using a SRBF surrogate technique.

6. Calculate failure probability P_F for iteration k using MCS.

7. Check the convergence criterion. If the convergence criterion is satisfied, stop; otherwise go to Step 8. In this study the convergence criterion is that the relative error of the failure probability P_F between two successive iterations is less than the tolerance. A tolerance value of 1% was applied in this study. For practical applications, another convergence criterion may be defined, e.g., the maximum number of response simulations has been reached. This will help control the total number of iterations performed in the reliability analysis.

8. Generate additional sample set with m sample points; set the iteration number $k = k + 1$.

9. Evaluate limit state function $g(\mathrm{x})$ for the additional sample set m generated in Step 8, then go to Step 4.

7. Numerical examples

Four numerical examples were solved using the proposed reliability analysis method. These include both mathematical and engineering problems found in literature. In this study, the proposed method based on three SRBFs, i.e., SRBF-MQ-LP, SRBF-CS20-LP, and SRBF-CS30-LP, is referred to as the SRBF-based MCS. The Direct MCS refers to MCS without using surrogate models. In the Direct MCS, the number of response simulations was determined by the MCS sample size. However, in the SRBF-based MCS, the number of response simulations was based on the surrogate modeling sample size. A total of $N = 10^6$ samples was adopted in MCS when surrogate models were used.

7.1 Example 1: a nonlinear limit state function

A nonlinear limit state function was studied in literature, as [21, 49, 50]:

$$g(\mathrm{x}) = exp\,(0.2x_1 + 6.2) - exp\,(0.47x_2 + 5.0) \quad (25)$$

where x_1 and x_2 are independent random variables following standard normal distributions (mean = 0; standard deviation = 1). The failure probability P_F = 0.009372 was obtained based on Direct MCS and used to compare with other solutions. The RBF surrogate models were constructed using the two variables sampled in the range of −3.0 to 3.0. All three surrogate models started with 10 sample points in the first iteration. With 10 sample points, the error of the estimated failure probability was 7.0, 3.0, and 1.8% for SRBF-MQ-LP, SRBF-CS20-LP, and SRBF-CS30-LP, respectively. In each subsequent iteration 10 more sample points were added. At convergence, the accuracy of SRBF models was improved; the error was reduced to 0.9, 0.8, and 1.3% for SRBF-MQ-LP, SRBF-CS20-LP, and SRBF-CS30-LP, respectively. Adequate accuracy of reliability analysis was achieved for all three SRBF surrogate models. The failure probability values obtained based on three surrogate models and the associated errors as compared to the solution obtained using Direct MCS are listed in **Table 2**. It took 4, 3, and 2 iterations for SRBF-MQ-LP, SRBF-CS20-, and SRBF-CS30-LP methods to converge, corresponding to 40, 30, and 20 sample points, respectively. A total of 40, 30, and 20 function evaluations (original limit state function) were required for the three SRBF-based MCS, respectively.

Method	Failure probability	Error (%)	No. of function evaluation
Direct MCS	0.009372		1,000,000
SRBF-MQ-LP	0.009456	0.9%	40
SRBF-CS20-LP	0.009443	0.8%	30
SRBF-CS30-LP	0.009498	1.3%	20

Table 2.
Example 1: numerical results.

7.2 Example 2: a cantilever beam

The reliability analysis of a cantilever beam with a concentrated load is conducted in this example [50]. The beam has a rectangular cross section. The performance requirement is the displacement at tip should be <0.15 in. Therefore, the limit state function is.

Random variable	Unit	Mean	Standard deviation	Distribution type
l	(in)	30	3.0	Lognormal
b	(in)	0.8359	0.08	Gaussian
h	(in)	2.5093	0.25	Lognormal

Table 3.
Example 2: random variables [50].

Method	Failure probability	Error (%)	No. of function evaluation
Direct MCS	0.02823		1,000,000
SRBF-MQ-LP	0.02550	9.7%	50
SRBF-CS20-LP	0.02814	0.3%	80
SRBF-CS30-LP	0.02774	1.7%	60

Table 4.
Example 2: numerical results.

$$g(l, b, h) = 0.15 - \frac{4Pl^3}{Ebh^3} \tag{26}$$

where P is the concentrated load, l is the beam length, b and h are the width and depth of the beam cross-section, and $E = 10^7$ psi is the Young's modulus. In this example $P = 80$ lb. was considered. **Table 3** lists the three random variables in this problem, i.e., l, b, and h.

All three SRBF surrogate models started with 20 sample points in the first iteration, with 10 more samples generated in each following iteration. The reliability analysis results and the corresponding sample sizes required for SRBF surrogate models were examined, as listed in **Table 4**. The failure probability estimated based on Direct MCS using Eq. (26) was 0.02823, which was regarded as the actual solution. It took 4, 7, and 5 iterations for SRBF-MQ-LP, SRBF-CS20-LP, and SRBF-CS30-LP to converge, respectively. With the initial 20 samples, the error of the estimated failure probability was 35.9, 19.4, and 9.7% for SRBF-MQ-LP, SRBF-CS20-LP, and SRBF-CS30-LP, respectively. With 50, 80, and 60 sample points, the error was reduced to 9.7% for SRBF-MQ-LP, 0.3% for SRBF-CS20-LP, and 1.7% for SRBF-CS30-LP. The errors in estimating the failure probability by SRBF surrogate models decreased as the sample size increased. The SRBF-MQ-LP model did not produce as accurate estimation of P_F as SRBF-CS20-LP and SRBF-CS30-LP, when the same sample size was used. In all three SRBF surrogate models, SRBF-CS20-LP provided the most accurate estimate of P_F, and the surrogate model SRBF-MQ-LP did not converge close to the actual solution. In this example, 60–80 sample points were required for SRBF-CS20-LP and SRBF-CS30-LP to achieve reasonably accurate surrogate models and estimates of the failure probability.

7.3 Example 3: a reinforced concrete beam section

This example is the reliability analysis of a singly-reinforced concrete beam section [51, 70]. Based on static equilibrium, the following nonlinear limit state function can be developed, as:

$$g(\mathbf{x}) = x_1 x_2 x_3 - x_4 \frac{x_1^2 x_2^2}{x_5 x_6} - M_n \tag{27}$$

Eq. (27) included six independent random variables: x_1 is the total cross-sectional area of rebars, x_2 is the yield strength of rebars, x_3 is the effective depth of section, x_4 is a dimensionless factor related to concrete stress-strain curve, x_5 is the compressive strength of concrete, and x_6 is the width of the concrete section. The limit state was for the ultimate bending moment strength of the section, and a bending moment limit $M_n = 211.20 \times 10^6$ N-mm was adopted in this study. **Table 5** lists the six input random variables and their statistical properties.

To start the reliability analysis, 30 sample points were used in the first iteration of all three SRBF surrogate models, and 10 additional samples were included in each subsequent iteration. **Table 6** lists the failure probability P_F values obtained using different methods, in addition to the required number of original function evaluations, representing the associated computational effort. Compared with P_F = 0.01102 obtained by Direct MCS, the errors of SRBF-MQ-LP, SRBF-CS20-LP and SRBF-CS30-LP were 0.8, 1.1, and 0.9%, respectively.

Figure 2 is the plot showing failure probability estimation versus sample size. All three SRBF models worked well and smooth convergence histories can be observed. The three SRBF models produced similar failure probabilities. The results by SRBF-CS20-LP and SRBF-CS30-LP were shown to be better than that using SRBF-MQ-LP when the sample size was small. Among the three SRBF models, SRBF-CS30-LP generated the most accurate approximation with the same sample size. As expected, more sample points resulted in reduced SRBF approximation errors. With the increase of the number of sample points or function evaluations (i.e., computational effort), a reduction in estimation error of the failure probability using the proposed SRBF models was observed. For example, the estimation error of P_F was reduced from 10.7 to 0.8% for SRBF-MQ-LP, 4.9–1.1% for SRBF-CS20-LP, and 4.1–0.9% for

Random variable	Unit	Mean	Standard deviation	Distribution type
x_1	(mm^2)	1,260	63	Lognormal
x_2	(N/mm^2)	250	17.5	Lognormal
x_3	(mm)	770	10	Lognormal
x_4	(N/A)	0.55	0.055	Lognormal
x_5	(N/mm^2)	30	4.5	Lognormal
x_6	(mm)	250	5	Lognormal

Table 5.
Example 3: random variables [70].

Method	Failure probability	Error (%)	No. of function evaluation
Direct MCS	0.1102		1,000,000
SRBF-MQ-LP	0.1110	0.8%	60
SRBF-CS20-LP	0.1114	1.1%	50
SRBF-CS30-LP	0.1112	0.9%	50

Table 6.
Example 3: numerical results.

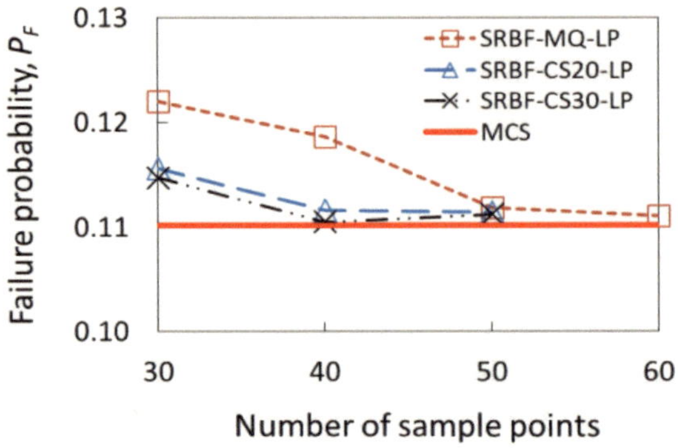

Figure 2.
Example 3: failure probability iterations.

SRBF-CS30-LP, respectively. SRBF-CS20-LP and SRBF-CS30-LP created with 40 samples and SRBF-MQ-LP created with 50 samples could provide fairly accurate reliability analysis results (<2% error of P_F).

7.4 Example 4: burst margin of a rotating disk

This example is the reliability analysis of a disk with an angular velocity of ω, as shown in **Figure 3** [50, 51]. The inner and outer radii of disk are R_i and R_o, respectively. The burst margin, M_b, of the disk refers to the safety margin before overstressing the disk, which is expressed as:

$$M_b(\alpha_m, S_u, \rho, \omega, R_o, R_i) = \sqrt{\frac{\alpha_m S_u}{\left(\frac{\rho\left(\frac{2\omega\pi}{60}\right)^2 \left(R_o^3 - R_i^3\right)}{3(385.82)(R_o - R_i)}\right)}} \tag{28}$$

If a lower bound value of 0.37473 is used, the limit state function of M_b can be written as:

$$g(\mathbf{x}) = M_b(\alpha_m, S_u, \rho, \omega, R_o, R_i) - 0.37473 \tag{29}$$

where S_u is the ultimate material strength, α_m is a dimensionless material utilization factor, and ρ is the mass density of material. **Table 7** lists the six random variables used in the example.

Similar as Example 3, all three surrogate models started with 30 sample points. In each subsequent iteration, 10 sample points were added. **Table 8** lists the

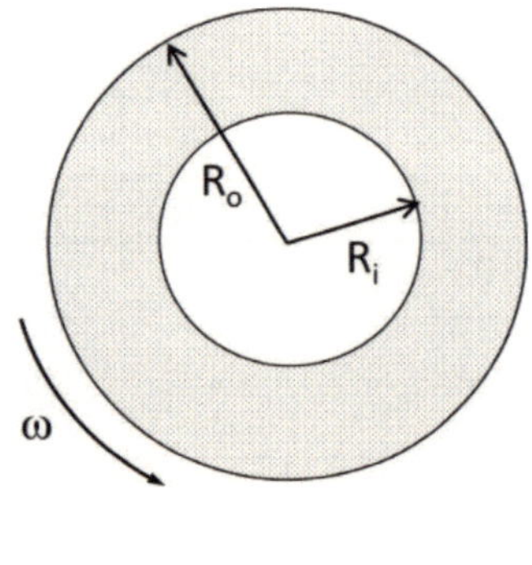

Figure 3.
Example 4: a rotating disk.

Random variable	Unit	Mean	Standard deviation	Distribution type
α_m	(N/A)	0.9378	0.04655	Weibull [a]
S_u	(lb/in^2)	220,000	5,000	Gaussian
ρ	(lb/in^3)	0.29	0.00577	Uniform [b]
ω	(rmp)	21,000	1,000	Gaussian
R_o	(in)	24	0.5	Gaussian
R_i	(in)	8	0.3	Gaussian

[a] Scale parameter = 25.508; shape parameter = 0.958.
[b] Uniformly distributed over 0.28-0.30.

Table 7.
Example 4: random variables [50, 51].

Method	Failure probability	Error (%)	No. of function evaluation
Direct MCS	0.001010		1,000,000
SRBF-MQ-LP	0.001067	5.6%	70
SRBF-CS20-LP	0.001018	0.8%	60
SRBF-CS30-LP	0.001005	0.5%	70

Table 8.
Example 4: numerical results.

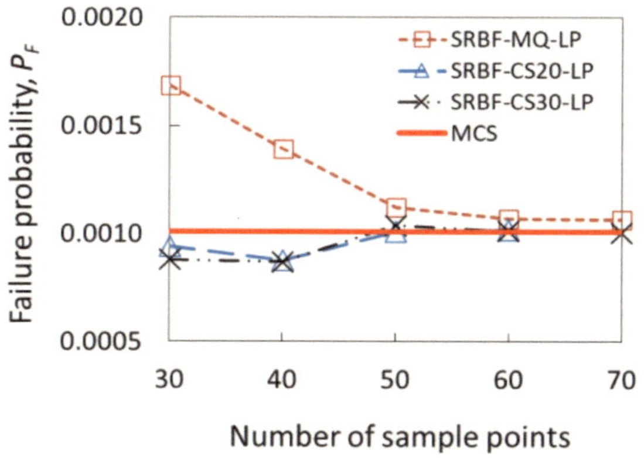

Figure 4.
Example 4: failure probability iterations.

estimated failure probability P_F in this study based on different SRBF surrogate models and the associated errors as compared to the solution obtained using Direct MCS. The augmented SRBF-based methods required 60–70 original function evaluations to converge. **Figure 4** illustrates the variation of the failure probability P_F versus number of sample points. In general with the increase of the sample size, a reduction was observed in the estimation errors of the failure probability P_F, from 67.1, 6.6, and 12.8% when 30 sample points were used, to 5.6, 0.8, and 0.5% at convergence for SRBF-MQ-LP, SRBF-CS20-LP, and SRBF-CS30-LP, respectively. The reliability analysis results based on surrogate models SRBF-CS20-LP and SRBF-CS30-LP were shown to be better that using SRBF-MQ-LP. It showed that with around 50 sample points very accurate SRBF-CS20-LP and SRBF-CS30-LP surrogate models could be created for reliability analysis.

8. Concluding remarks

Augmented RBFs are suitable for creating accurate surrogate models for linear and nonlinear responses. When combined with a sampling method such as MCS, they can be used in reliability analysis and provide accurate estimation of the failure probability. In spite of their excellent model accuracy, the most appropriate number of sample points is not known beforehand. To provide an improved and automated approach using the RBF surrogate models in reliability analysis, a SRBF surrogate modeling technique was developed and tested in this study, so that the RBF surrogate models could be used in an iterative yet efficient manner. In this chapter, three augmented RBFs, including multiquadric function and two compactly supported basis functions were considered. To evaluate the proposed SRBF surrogate modeling method for reliability analysis, its numerical accuracy and computational efficiency was examined.

Numerical examples including existing mathematical and engineering problems were studied using the proposed method. Accurate failure probability results were achieved using a reasonable sample size within a few iterations. The required number of response simulations or function evaluations was relatively small. All three SRBF models produced similar accuracy, and the surrogate models based on SRBF-CS20-LP and SRBF-CS30-LP produced more accurate reliability analysis results, especially when a smaller sample size was adopted. This study shows that the proposed reliability analysis method is efficient and has a promising potential for application to complex engineering problems involving expensive simulations. Further research includes efficient sequential sampling methods that can be combined with the SRBF methods, and the optimal approach to determine the sample sizes used in each iteration of the SRBF methods.

Author details

Qian Wang
Department of Civil and Environmental Engineering, Manhattan College, Riverdale, NY, USA

*Address all correspondence to: qian.wang@manhattan.edu

References

[1] Ang AHS, Tang WH. Probability Concepts in Engineering Planning and Design. Vol. 1 Basic Principles. New York: Wiley; 1975

[2] Madsen HO, Krenk S, Lind NC. Methods of Structural Safety. Englewood Cliffs, NJ: Printice-Hall; 1986

[3] Ditlevsen O, Madsen HO. Structural Reliability Methods. Chichester: Wiley; 1996

[4] Hasofer AM, Lind NC. Exact and invariant second moment code format. Journal of Engineering Mechanics. 1974; **100**(1):111-121

[5] Kiureghian D, Lin H-Z, Hwang S-J. Second order reliability approximations. Journal of Engineering Mechanics, ASCE. 1987;**113**(8):1208-1225

[6] Hohenbichler M, Gollwitzer S, Kruse W, Rackwitz R. New light on first and second-order reliability methods. Structural Safety. 1987;**4**:267-284

[7] Low BK, Tang WH. Efficient spreadsheet algorithm for first-order reliability method. Journal of Engineering Mechanics, ASCE. 2007; **133**(12):1378-1387

[8] Rubinstein RY. Simulation and the Monte Carlo Method. New York: Wiley; 1981

[9] Melchers RE. Importance sampling in structural systems. Structural Safety. 1989;**6**(1):3-10

[10] Au SK, Beck JL. Estimation of small failure probabilities in high dimensions by subset simulation. Probabilistic Engineering Mechanics. 2001;**16**(4): 263-277

[11] Brown SA, Sepulveda AE. Approximation of system reliability using a shooting Monte Carlo approach. AIAA Journal. 1997;**35**(6):1064-1071

[12] Au SK, Wang Y. Engineering Risk Assessment with Subset Simulation. New York: John Wiley & Sons, Inc.; 2014

[13] Bucher C, Most T. A comparison of approximate response functions in structural reliability analysis. Probabilistic Engineering Mechanics. 2008;**23**:154-163

[14] Bai YC, Han X, Jiang C, Liu J. Comparative study of surrogate modeling techniques for reliability analysis using evidence theory. Advances in Engineering Software. 2012;**53**:61-71

[15] Wong FS. Slope reliability and response surface method. Journal of Geotechnical Engineering, ASCE. 1985; **111**(1):32-53

[16] Faravelli L. Response surface approach for reliability analysis. Journal of Engineering Mechanics, ASCE. 1989; **115**(12):2763-2781

[17] Bucher CG, Bourgund U. A fast and efficient response surface approach for structural reliability problems. Structural Safety. 1990;**7**(1):57-66

[18] Rajashekhar MR, Ellingwood BR. A new look at the response surface approach for reliability analysis. Structural Safety. 1993;**12**(3):205-220

[19] Guan XL, Melchers RE. Multitangent-plane surface method for reliability calculation. Journal of Engineering Mechanics, ASCE. 1997; **123**(10):996-1002

[20] Das PK, Zheng Y. Cumulative formation of response surface and its use in reliability analysis. Probabilistic

Engineering Mechanics. 2000;**15**(4): 309-315

[21] Kmiecik M, Guedes Soares C. Response surface approach to the probability distribution of the strength of compressed plates. Marine Structures. 2002;**15**(2):139-156

[22] Romero VJ, Swiler LP, Giunta AA. Construction of response surfaces based on progressive-lattice-sampling experimental designs with application to uncertainty propagation. Structural Safety. 2004;**26**(2):201-219

[23] Mollon G, Daniel D, Abdul HS. Probabilistic analysis of circular tunnels in homogeneous soil using response surface methodology. Journal of Geotechnical and Geoenvironmental Engineering. 2009; **135**(9):1314-1325

[24] Lv Q, Sun HY, Low BK. Reliability analysis of ground-support interaction in circular tunnels using response surface method. International Journal of Rock Mechanics and Mining Sciences. 2011;**48**(8):1329-1343

[25] Lv Q, Low BK. Probabilistic analysis of underground rock excavations using response surface method and SORM. Computers and Geotechnics. 2011;**38**: 1008-1021

[26] Kim SH, Na SW. Response surface method using vector projected sampling points. Structural Safety. 1997;**19**(1): 3-19

[27] Gayton N, Bourinet JM, Lemaire M. CQ2RS: A new statistical approach to the response surface method for reliability analysis. Structural Safety. 2003;**25**(1):99-121

[28] Zheng Y, Das PK. Improved response surface method and its application to stiffened plate reliability analysis. Engineering Structures. 2000; **22**(5):544-551

[29] Gavin HP, Yau SC. High-order limit state functions in the response surface method for structural reliability analysis. Structural Safety. 2008;**30**(2): 162-179

[30] Liu YW, Moses F. A sequential response surface method and its application in the reliability analysis of aircraft structural systems. Structural Safety. 1994;**16**:39-46

[31] Kang S-C, Koh H-M, Choo JF. An efficient response surface method using moving least square approximation for structural reliability analysis. Probabilistic Engineering Mechanics. 2010;**25**:365-371

[32] Papadrakakis M, Papadopoulos V, Lagaros ND. Structural reliability analysis of elastic–plastic structures using neural networks and Monte Carlo simulation. Computer Methods in Applied Mechanics and Engineering. 1996;**136**(1–2):145-163

[33] Hurtado JE, Alvarez DA. Neural-network-based reliability analysis: A comparative study. Computer Methods in Applied Mechanics and Engineering. 2001;**191**(1–2):113-132

[34] Cardoso JB, Almeida JR, Dias JM, Coelho PG. Structural reliability analysis using Monte Carlo simulation and neural networks. Advances in Engineering Software. 2008;**39**(6): 505-513

[35] Papadopoulos V, Giovanis DG, Lagaros ND, Papadrakakis M. Accelerated subset simulation with neural networks for reliability analysis. Computer Methods in Applied Mechanics and Engineering. 2012; **223–224**:70-80

[36] Gomes HM, Awruch AM. Comparison of response surface and neural network with other methods for structural reliability analysis. Structural Safety. 2004;**26**:49-67

[37] Dai HZ, Zhao W, Wang W, Cao ZG. An improved radial basis function network for structural reliability analysis. Journal of Mechanical Science and Technology. 2011;**25**(9):2151-2159

[38] Simpson TW, Mauery TM, Korte JJ, Mistree F. Kriging surrogate models for global approximation in simulation-based multidisciplinary design optimization. AIAA Journal. 2001; **39**(12):2233-2241

[39] Kaymaz I. Application of Kriging method to structural reliability problems. Structural Safety. 2005;**27**(2): 133-151

[40] Bichon BJ, Eldred MS, Swiler LP, Mahadevan S, McFarland JM. Efficient global reliability analysis for nonlinear implicit performance functions. AIAA Journal. 2008;**46**(10):2459-2468

[41] Echard B, Gayton N, Lemaire M. AK-MCS: An active learning reliability method combining Kriging and Monte Carlo simulation. Structural Safety. 2011;**33**(2):145-154

[42] Zhang J, Zhang L, Tang W. Kriging numerical models for geotechnical reliability analysis. Soils and Foundations. 2011;**51**(6):1169-1177

[43] Echard B, Gayton N, Lemaire M, Relun N. A combined importance sampling and Kriging reliability method for small failure probabilities with time-demanding numerical models. Reliability Engineering & System Safety. 2013;**111**:232-240

[44] Dubourg V, Sudret B, Deheeger F. Metamodel-based importance sampling for structural reliability analysis. Probabilistic Engineering Mechanics. 2013;**33**:47-57

[45] Gaspar B, Teixeira AP, Guedes Soares C. Assessment of the efficiency of Kriging surrogate models for structural reliability analysis. Probabilistic Engineering Mechanics. 2014;**37**:24-34

[46] Yun W, Lu Z, Jiang X. An efficient reliability analysis method combining adaptive Kriging and modified importance sampling for small failure probability. Structural and Multidisciplinary Optimization. 2018; **58**:1383-1393

[47] Tunga MA, Demiralp M. A factorized high dimensional model representation on the nodes of a finite hyperprismatic regular grid. Applied Mathematics and Computation. 2005; **164**(3):865-883

[48] Tunga MA, Demiralp M. Hybrid high dimensional model representation (HHDMR) on the partitioned data. Journal of Computational and Applied Mathematics. 2006;**185**(1): 107-132

[49] Chowdhury R, Rao BN, Prasad AM. High-dimensional model representation for structural reliability analysis. Communications in Numerical Methods in Engineering. 2009;**25**:301-337

[50] Chowdhury R, Rao BN. Assessment of high dimensional model representation techniques for reliability analysis. Probabilistic Engineering Mechanics. 2009;**24**:100-115

[51] Rao BN, Chowdhury R. Enhanced high-dimensional model representation for reliability analysis. International Journal for Numerical Methods in Engineering. 2009;**77**:719-750

[52] Suykens JAK, Vandewalle J. Least squares support vector machine classifiers. Neural Processing Letters. 1999;**9**(3):293-300

[53] Zhao H, Ru Z, Chang X, Yin S, Li S. Reliability analysis of tunnel using least square support vector machine. Tunnelling and Underground Space Technology. 2014;**41**:14-23

[54] Zhao H. Slope reliability analysis using a support vector machine. Computers and Geotechnics. 2008;**35**: 459-467

[55] Hurtado JE. Filtered importance sampling with support vector margin: A powerful method for structural reliability analysis. Structural Safety. 2007;**29**(1):2-15

[56] Bourinet J-M, Deheeger F, Lemaire M. Assessing small failure probabilities by combined subset simulation and support vector machines. Structural Safety. 2011;**33**(6):343-353

[57] Tan XH, Bi WH, Hou XL, Wang W. Reliability analysis using radial basis function networks and support vector machines. Computers and Geotechnics. 2011;**38**(2):178-186

[58] Krishnamurthy T. Response Surface Approximation with Augmented and Compactly Supported Radial Basis Functions. Technical Report AIAA-2003-1748. Reston, VA: AIAA; 2003

[59] Goel T, Haftka RT, Shyy W, Queipo NV. Ensemble of surrogates. Structural and Multidisciplinary Optimization. 2007;**33**(3):199-216

[60] Acar E, Rais-Rohani M. Ensemble of metamodels with optimized weight factors. Structural and Multidisciplinary Optimization. 2009;**37**(3):279-294

[61] Yin H, Fang H, Wen G, Gutowski M, Xiao Y. On the ensemble of metamodels with multiple regional optimized weight factors. Structural and Multidisciplinary Optimization. 2018; **58**:245-263

[62] Ye P, Pan G, Dong Z. Ensemble of surrogate based global optimization methods using hierarchical design space reduction. Structural and Multidisciplinary Optimization. 2018; **58**:537-554

[63] Fang H, Rais-Rohani M, Liu Z, Horstemeyer MF. A comparative study of surrogate modeling methods for multi-objective crashworthiness optimization. Computers & Structures. 2005;**83**(25–26):2121-2136

[64] Wang Q, Fang H, Shen L. Reliability analysis of tunnels using a meta-modeling technique based on augmented radial basis functions. Tunnelling and Underground Space Technology. 2016;**56**:45-53

[65] Fang H, Horstemeyer MF. Global response approximation with radial basis functions. Engineering Optimization. 2006;**38**(04):407-424

[66] Fang H, Wang Q. On the effectiveness of assessing model accuracy at design points for radial basis functions. Communications in Numerical Methods in Engineering. 2008;**24**(3):219-235

[67] Wu Z. Compactly supported positive definite radial function. Advances in Computational Mathematics. 1995;**4**:283-292

[68] Montgomery DC. Design and Analysis of Experiments. New York: John Wiley & Sons, Inc.; 2001

[69] Taguchi G. Taguchi Method-Design of Experiments, Quality Engineering Series. Vol. 4. Tokyo: ASI Press; 1993

[70] Zhou J, Nowak AS. Integration formulas to evaluate functions of random variables. Structural Safety. 1988;**5**(4):267-284

Chapter 5

Reliability of Microelectromechanical Systems Devices

Wu Zhou, Jiangbo He, Peng Peng, Lili Chen and Kaicong Cao

Abstract

Microelectromechanical systems (MEMS) reliability issues, apart from traditional failure mechanisms like fatigue, wear, creep, and contamination, often involve many other specific mechanisms which do not damage the system's function but may degrade the performance of MEMS devices. This chapter focuses on the underlying mechanisms of specific reliability issues, storage long-term drift and thermal drift. The comb finger capacitive micro-accelerometers are selected as the case for this study. The material viscoelasticity of packaging adhesive and thermal effects induced by structure layout are considered so as to explain the physical phenomenon of output change over time and temperature. Each section showcases the corresponding experiments and analysis of reliability.

Keywords: MEMS reliability, micro-accelerometer, drift, dielectric charging, viscoelasticity

1. Introduction

Microelectromechanical systems technology has been widely applied in areas such as inertial navigation, RF/microwave communications, optical communications, energy resources, biomedical engineering, environmental protection, and so on. The MEMS-related products involve micro-accelerometers, gyroscopes, micro-resonators, microswitches, micro-pumps, pressure sensors, energy harvesters, etc. Many new designs and prototypes of MEMS are produced and marketed in large numbers year by year. Only a few, however, can be used as mature products in the field with requirements for high performance. The main obstacle is that the reliability issues of micro systems involve numerous physical failure mechanisms covering the aspects of mechanical structures, electrical components, and packaging and working conditions [1–6].

The industrial standards for MEMS reliability, so far, are not existent because the behavior of MEMS is highly dependent on the designs and fabrication of specific micro systems. This is attributed to the complication and diversity of micro-devices. The coupling effects between different physical domains add much more complexities to the analysis of failure modes. For example, the effects of thermal expansion are not only determined by the difference of coefficients of thermal expansion (CTE) but are equally highly impacted by the structural layout [7, 8]. A failure mode, therefore, can exhibit many different reliability phenomena in different devices; meanwhile, the exhibited same phenomenon like drift and stability may not result from the same physical mechanism. Current publications on MEMS' reliability involved almost every aspect of micro systems including structures, electrical components, materials, electronics, packaging, and so on. Performed literature research reveals a wide coverage of topics, ranging from basic physical mechanisms to engineering applications and from single structural units to entire device systems. Compared with the failure modes of mechanical structures and electrical components which have a certain similarity to macro systems [9–17], the reliability issue of systematical behaviors is significantly more important because it always originates from the interaction between its components or sub-systems which by themselves can work normally [18–20].

In this chapter, the typical reliability issue regarding the MEMS packaging effects of micro-accelerometer, selected as a specific device, is concerned. MEMS packaging, developed from integrated circuit (IC) packaging, is to integrate the fabricated device and circuit. Yet, the two packaging technologies are significantly different. The functions of IC packaging are mainly to protect, power, and cool the microelectronic chips or components and provide electrical and mechanical connection between the microelectronic part and the outside world [21]. Packaging of MEMS is much more complex than that for the IC due to the inherent complexities in structures and intended performances. Many MEMS products involve precision movement of solid components and need to interface with different outside environments, the latter being determined by their specific functions of biological, chemical, electromechanical, and optical nature. Therefore, MEMS packaging processes have to provide more functionalities including better mechanical protection, thermal management, hermetic sealing, complex electricity, and signal distribution [22].

A schematic illustration of a typical MEMS packaging is shown in **Figure 1**. Both the MEMS sensor die and the application specific integrated circuit (ASIC) are mounted onto a substrate using a die attach adhesive. The sensor die is covered by a MEMS cap in order to prevent any particles to ruin the sensitive part. Thereafter, they are wire bonded to acquire the electric connection and enclosed by the molding compound to provide protection from mechanical or environmentally induced damages [12].

Packaging, in particular, is an integral part of the MEMS design and plays a crucial role in the device stability. Package-induced stresses appear to be unavoidable in almost all MEMS components due to CTEs' (Coefficient of Thermal Expansion) mismatch of the packaging materials during the packaging process, especially in the die bonding and sealing process. The stresses existing in structures and interfaces form a stable equilibrium of micro-devices based on deformation compatibility conditions [23]. The formed equilibrium, however, is prone to be upset by the shift of material properties and/or structure expansion induced by the temperature load. The material aspects are always related to the packaging adhesive

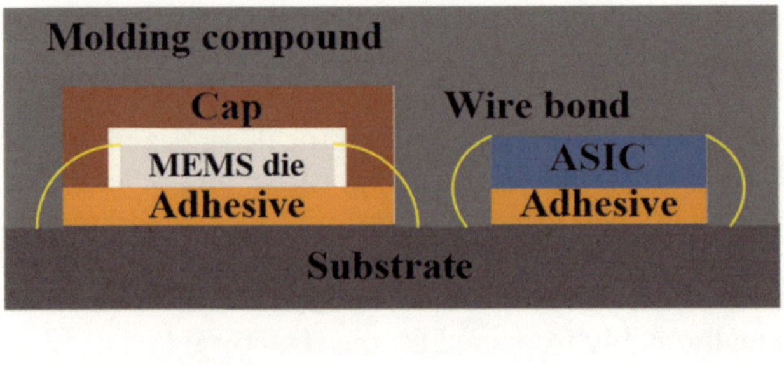

Figure 1.
A schematic illustration of typical MEMS packaging.

because silicon, glass, and ceramic exhibit an excellent stable property. This adhesive however, a polymer-based material, is often simply assumed to be linear elastic [24–26]. This assumption could give a relatively accurate evaluation of device performance in the low- and medium-precision application fields, but could not be used to predict the long-term stability or drift in areas requiring high precision without taking in consideration the viscoelasticity of polymers [27]. This chapter will deal with the stability regarding the viscoelasticity of packaging materials. With regard to the structure aspects, the main reliability issue is the thermal effects induced by the temperature change. The level of effects is attributed to the range of thermal mismatch and the structural layout. The former is unchangeable for a specific device because the structural materials are readily selected, while the latter, although of interest, lacks to attract further research, because researchers preferred a temperature compensation by external components than search for the underlying mechanism of thermal effects of devices. The current compensation technology can be categorized into active compensation and passive compensation.

1.1 Active compensation

Active compensation requires a temperature sensor to measure the device operating temperature, which is then fed back to a controller to keep the environment temperature constant. This is achieved by means of an algorithm and a thermometer, so as to control and compensate the output offset induced by temperature change.

The temperature control is the most widely applied technology regarding active compensation. However, the micro-oven may be regarded as a disadvantage of this technology, because it makes the device much bigger. In order to overcome this, Xu et al. [28] proposed a miniaturized and integrated heater that enables low power. Besides temperature control, modification of the performance is another broadly used compensation technology. For the MEMS sensor, its thermal drifts, such as thermal drift of bias (TDB) and thermal drift of scale factor (TDSF), are usually tested and recorded firstly. Then, when the MEMS sensor is in operation, the output is modified mathematically based on the recorded thermal drifts. For the MEMS resonator, the frequency modification by electrostatic stiffness is a frequently used technology [29]. In this technology, the temperature is fed back to control the operating voltage of the MEMS resonator and then change the electrostatic stiffness and consequently the frequency.

The position of the temperature sensor is critical for the compensation technology of the modification of the performance. In many MEMS devices, the temperature is integrated in the ASIC die, and the ASIC die is integrated with the MEMS die through the package. As such, the temperature sensors actually measure the temperature of the ASIC die. This, though, is error-prone regarding the temperature measurement of the MEMS die. In order to improve the temperature accuracy, several innovative technologies for temperature measuring have been proposed [30–32].

In order to compensate the thermal drift of frequency of MEMS resonator, Hopcroft et al. [33] extracted the temperature information from the variation of the quality factor. Kose et al. [34] reported a compensation method for a capacitive MEMS accelerometer by using a double-ended tuning fork resonator integrated with the accelerometer on the same die for measuring temperature. Du et al. [35] presented a real-time temperature compensation algorithm for a force-rebalanced MEMS capacitive accelerometer which relies on the linear relationship between the temperature and its dynamical resonant frequency.

1.2 Passive compensation

Active compensation is simpler and uncomplicated; however, it typically involves the additional circuitry and power consumption. On the contrary, the passive compensation does not need the additional circuitry and power consumption.

1.2.1 Passive compensation for TCEM

The current passive compensation technology for temperature coefficient of elastic modulus (TCEM) includes electrostatic stiffness modification, composite structure, and high doping. Melamud et al. [36] proposed a Si-SiO_2 composite resonator, as shown in **Figure 2a**. SiO_2 covers the surface of the silicon beam to form a composite resonator beam. Because TCEMs of silicon and SiO_2 are negative and positive, respectively, the Si-SiO_2 composite resonator can realize the passive compensation for the thermal drift of frequency. Liu et al. [37] also proposed an Al/SiO_2 composite MEMS resonator with the passive compensation ability for the thermal drift of frequency.

TCEM of single crystal silicon can also be compensated by high doping. A suggested mechanism is that heavy doping strains the crystal lattice and shifts the electronic energy bands, resulting in a flow of charge carriers to minimize the free energy, thereby changing the elastic properties [38]. Hajjam [39] reported a high phosphorus-doped silicon MEMS resonator with thermal drift of frequency down to 1.5 ppm/°C. Samarao [40] reported that MEMS resonator with high concentration

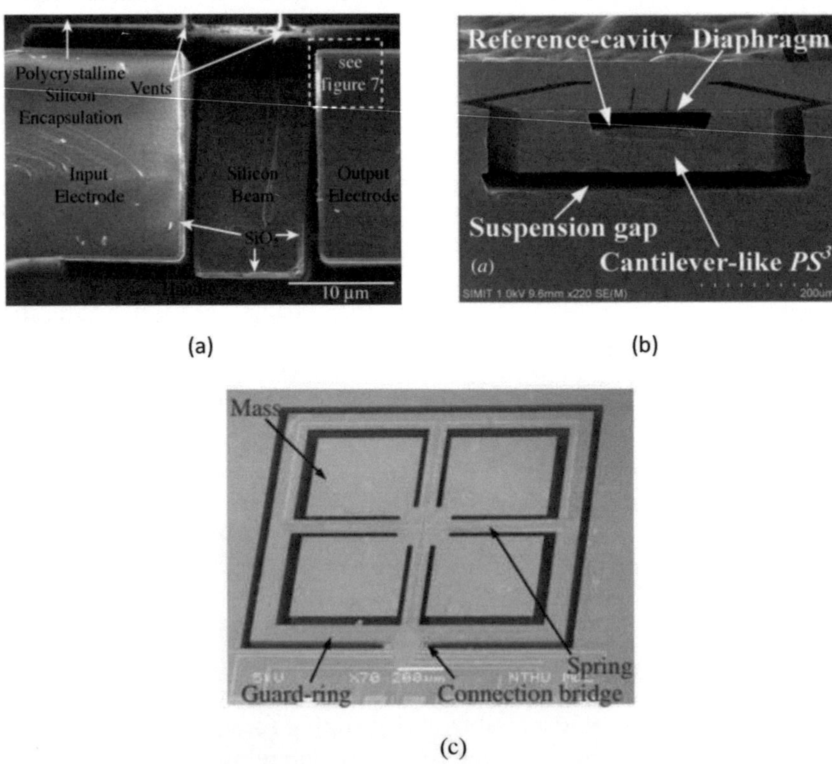

Figure 2.
MEMS devices with passive compensation, the ability of isolating the thermal stress. (a) Composite resonator [36], (b) pressure sensor isolating the thermal stress [43], and (c) three-axis piezo-resistive accelerometer pressure sensor isolating the thermal stress [44].

of boron doping and aluminum has thermal drifts of 1.5 and 2.7 ppm/°C, respectively.

The result of passive compensation depends on the precise structure design and is susceptible to fabrication error. As such, passive compensation generally cannot suppress the thermal drift fully. In several reports, active compensation and passive compensation are incorporated to suppress the thermal drift. For instance, Lee et al. [41] incorporate the electrostatic stiffness and Si-SiO_2 composite structure to compensate the thermal drift of the frequency of MEMS resonator. As such, 2.5 ppm drift of frequency over 90°C full scale is obtained.

1.2.2 Passive compensation for thermal stress/deformation

The improvement on structure design or package to suppress the thermal stress/deformation is an effective passive compensation technology for thermal stress/deformation.

The soft adhesive attaching, such as using rubber adhesive with an elastic modulus lower than 10M, is commonly employed to suppress thermal stress/deformation induced by the package [42]. Furthermore, the soft attaching area is also reduced to obtain lower thermal stress/deformation. Besides the passive compensation technology in packaging, improvement on structure design is also successfully employed to suppress thermal stress/deformation. Wang et al. [43] proposed a pressure sensor structure that can isolate stress, as shown in **Figure 2b**. They used cantilever beam to suspend the detection component of the pressure sensor, thereby isolating the influence of encapsulation effect on the sensor. Based on a floating ring, Hsieh et al. [44] suggested a three-axis piezo-resistive accelerometer with low thermal drift, as shown in **Figure 2c**.

1.2.3 Making the TCEM and thermal stress/deformation compensate each other

It is very promising to make the TCEM and thermal stress/deformation balance each other out. Hsu et al. [45] used thermal deformation to adjust the capacitance gap, so as to achieve the automatic adjustment of electrostatic stiffness and compensation for the variation of mechanical stiffness induced by temperature. Myers et al. [46] employed the thermal stress caused by the mismatch of CTE to compensate the frequency drift induced by TCEM. In this chapter, the thermal analysis is carried out in order to investigate the impacts of a structured layout of a sensing element on the drift over temperature of micro-accelerometers, and an optimized structure is proposed to improve the thermal stability.

2. Reliability analysis and experiments

2.1 Reliability regarding viscoelasticity

2.1.1 Polymer viscoelasticity

Viscoelasticity is a distinguishing characteristic of materials such as polymer. It exhibits both elastic and viscous behavior. The elasticity responding to stress is instantaneous, while the viscous response is time-dependent and varies with temperature. The viscoelastic behavior can be expressed with Hookean springs and Newtonian dashpot, which correspond to elastic and viscous properties, respectively. To measure the viscoelastic characteristics, stress relaxation or creep tests are often implemented. Stress relaxation of viscoelastic materials is commonly described using a generalized Maxwell model, which is shown in **Figure 3**. It consists of a number of springs and dashpots connected in parallel, which represent elasticity and viscosity, respectively.

The Maxwell model can be described as a Prony series, which can be expressed with Eqs. (1) and (2) as below [47]:

$$E(t) = \frac{\sigma(t)}{\varepsilon_0} = E_\infty + \sum_{i=1}^{N} E_i \exp\left(-\frac{t}{\tau_i}\right) \tag{1}$$

$$\tau_i = \frac{\eta_i}{E_i} \tag{2}$$

where $E_{(t)}$ is the relaxation modulus; $\sigma_{(t)}$ is the stress; ε_0 is the imposed constant strain; E_∞ is the fully relaxed modulus; E_i and τ_i are referred to as a Prony pair; E_i and η_i are the elastic modulus and viscosity of Prony pair; τ_i is the relaxation time of ith Prony pair; N is the number of Prony pairs.

The relaxation data, for normalization, can be modeled by a master curve, which translates the curve segments at different temperatures to a reference temperature with logarithmic coordinates according to a time-temperature superposition [48]. The master curve can be fitted by a third-order polynomial function, such as

$$\log a_T(T) = C_1(T - T_0) + C_2(T - T_0)^2 + C_3(T - T_0)^3 \tag{3}$$

where a_T is the offset values at different temperatures (T); C_1, C_2, and C_3 are constants; and T_0 is the reference temperature.

Taking an epoxy die adhesive used in a MEMS capacitive accelerometer as an example, a series of stress-relaxation tests were performed using dynamic mechanical analysis [49]. The experimental temperature range was set from 25 to 125°C with an increment of 10°C and an increase rate of 5°C/min, and at each test point, 5 min was allowed for temperature stabilization, and 0.1% strain was applied on the adhesive specimen for 20 min, followed by a 10 min recovery. The test results are shown in **Figure 4**.

The shift distance of individual relaxation curve with reference temperature of 25°C was shown in **Figure 5**. Then the three coefficients of the polynomial function (Eq. (3)) were determined to be $C_1 = 0.223439$, $C_2 = -0.00211$, and $C_3 = 5.31163 \times 10^{-6}$.

Subsequently the master curve can be acquired, and a Prony series having nine Prony pairs (Eq. (1)) was used to fit to the master curve, as shown in **Figure 6**. The coefficients of the Prony pairs are listed in **Table 1**, where E_0 is the instantaneous modulus when time is zero.

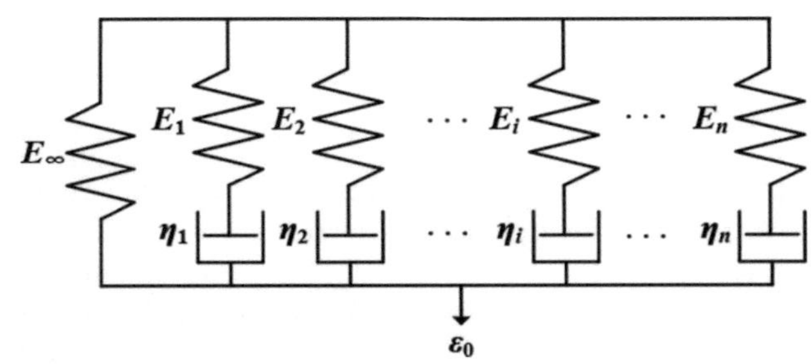

Figure 3.
Generalized Maxwell model to describe the viscoelastic behavior.

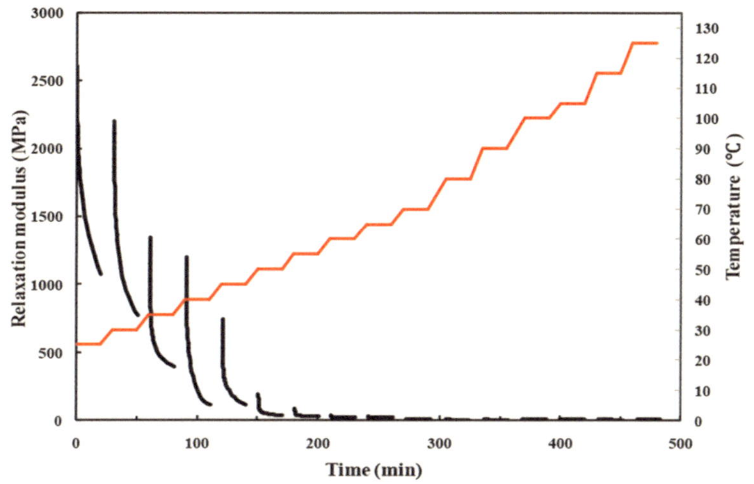

Figure 4.
Stress-relaxation test results of an epoxy die attach adhesive.

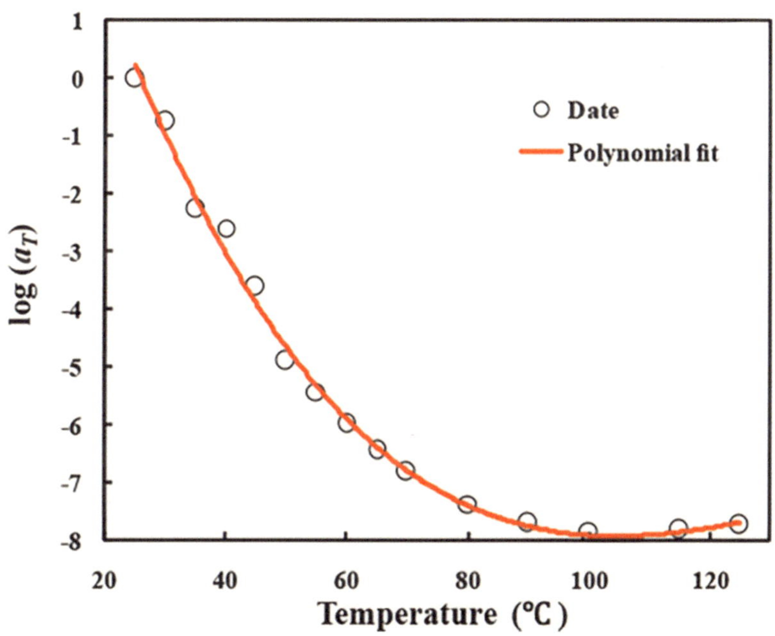

Figure 5.
Shift distance plot of individual relaxation curve with reference temperature of 25°C and polynomial fit.

2.1.2 Viscoelasticity-induced stability problem of MEMS

The viscoelasticity-related issue has become one of the most critical steps for assessing the packaging quality and output performance of highly precise MEMS sensors [50]. Applying the viscoelastic property to model the MEMS devices could yield a better agreement with the results observed in experiments than the previous elastic model [51]. The packaging stress in the MEMS was influenced not only by the temperature change but also by its change rate due to the time-dependent property of polymer adhesives [52]. Besides, the viscoelastic behavior influenced by moisture was recognized as the cause of the long-term stability of microsensors in storage [53].

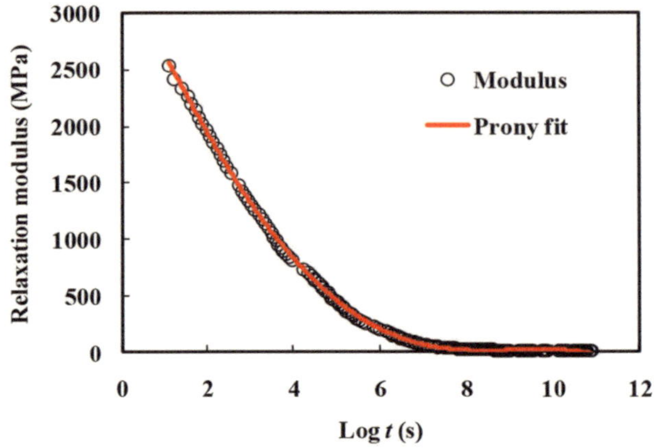

Figure 6.
Prony series fitted to master curve.

i	E_i/E_0	τ_i
1	0.08510	3041.87694
2	0.14589	981765.85865
3	0.22654	243.32699
4	0.11248	2083.25650
5	0.15906	48993.21024
6	0.21617	31.16988
7	0.03923	5052.23180
8	0.00025	6992.77554
9	0.00676	3321.33054

E_0 = 2744.76252 MPa.

Table 1.
Prony pairs of the die attach adhesive.

In the following, the output stability of a capacitive micro-accelerometer was investigated using both simulation and experimental methods. The simulation introduced the Prony series modulus into the whole finite element model (FEM) in Abaqus software to acquire the output of the micro-accelerometers over time and temperature. The thermal experiment was carried out in an incubator with an accurate temperature controller. The full loading history used in both simulation and the experiment is shown in **Figure 7**. The red-marked points represent the starting or ending points of a loading step. The bias and sensitivity of the accelerometers subjected to the thermal cycles are shown in **Figure 8**. The observed output drift in the simulation and the experiments indicates that the viscoelasticity of adhesive was the main cause of the deviation of zero offset and sensitivity. The underlying mechanism can be attributed to the time- and temperature-dependent stress and deformation states of the sensitive components of the micro-accelerometers.

It is evident that the output of the sensor after each thermal cycle will not change if the adhesive is assumed to be linear elastic.

The storage long-term drift of the accelerometer was also assessed by simulation and experimental methods based on the viscoelasticity of polymer adhesive. The residual stress formed in the curing process of the packaging would develop over time due to the internal strain changing with the relaxation of stress.

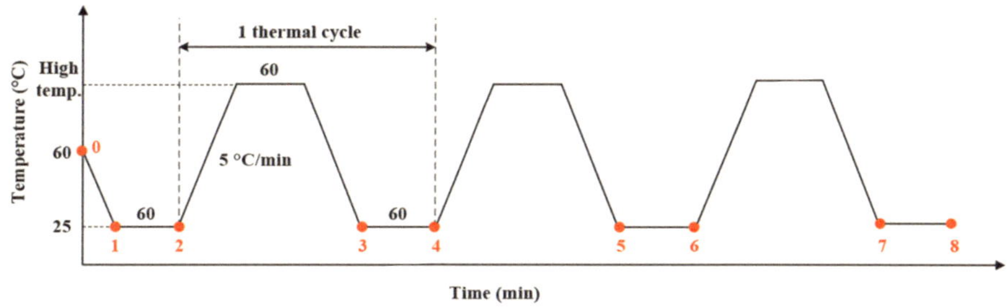

Figure 7.
Loading history for the analysis.

The temperature profile in the simulation started from 60 (curing temperature) to 25°C (room temperature), and then the sensor was kept at 25°C for 12 months. The variation of the bias of the sensor was shown in **Figure 9**. The bias decreased over time due to stress relaxation and declined by about 21 mg in 12 months. After the sensor was kept at 25°C for about 10 months, the bias reached a steady state (<1 mg per month). The variation trend of the bias is generally consistent with the master curve of the relaxation modulus (**Figure 5**), which indicates that the package-induced stress will gradually be released because of adhesive viscoelastic characteristic in the long-term storage period.

2.2 Thermal drift of MEMS devices

The thermal drift of MEMS devices is related to its material, structure, interface circuit, and so on. The temperature coefficient of elastic modulus (TCEM) and the thermal stress/deformation are the factors studied mostly.

2.2.1 TCEM

Due to the excellent mechanical property, single crystal silicon is suitable for high-performance sensors, oscillators, actuators, etc. However, the elastic modulus of single crystal silicon is temperature dependent. Because the single crystal silicon is anisotropic, the temperature behavior of elasticity is more properly described by the temperature coefficients of the individual components of the elasticity tensor, Tc11, Tc12, etc., as shown in **Table 2**. In order to simplify the designing, the value of TCEM for typical axial loading situations is usually employed and equal to approximately −64 ppm/°C at room temperature [54].

The performance of MEMS devices is influenced by TCEM through the stiffness. In fact, the temperature coefficient of stiffness (TCS) is the sum of TCEM and CTE (coefficient of thermal expansion). CTE is 2.6 ppm/°C at room temperature and much smaller than TCEM. Therefore, TCS is mainly determined by TCEM. The effect of TCEM on performance is dependent on the principle of the MEMS device. If the MEMS device is oscillating at a fixed frequency for time reference, sensing or generating Coriolis force, its frequency has a thermal drift of TCS/2, because the frequency is related to the square root of stiffness. On the other hand, the thermal drift of the MEMS device, which the mechanical deformation is employed for sensing, such as the capacitive sensor, is equal to TCS, because the performance is related to the stiffness [55].

2.2.2 Thermal stress/deformation

Single crystal silicon expands with temperature and has a CTE of 2.6 ppm/°C at room temperature. The expansion induced by CTE can cause the variation of

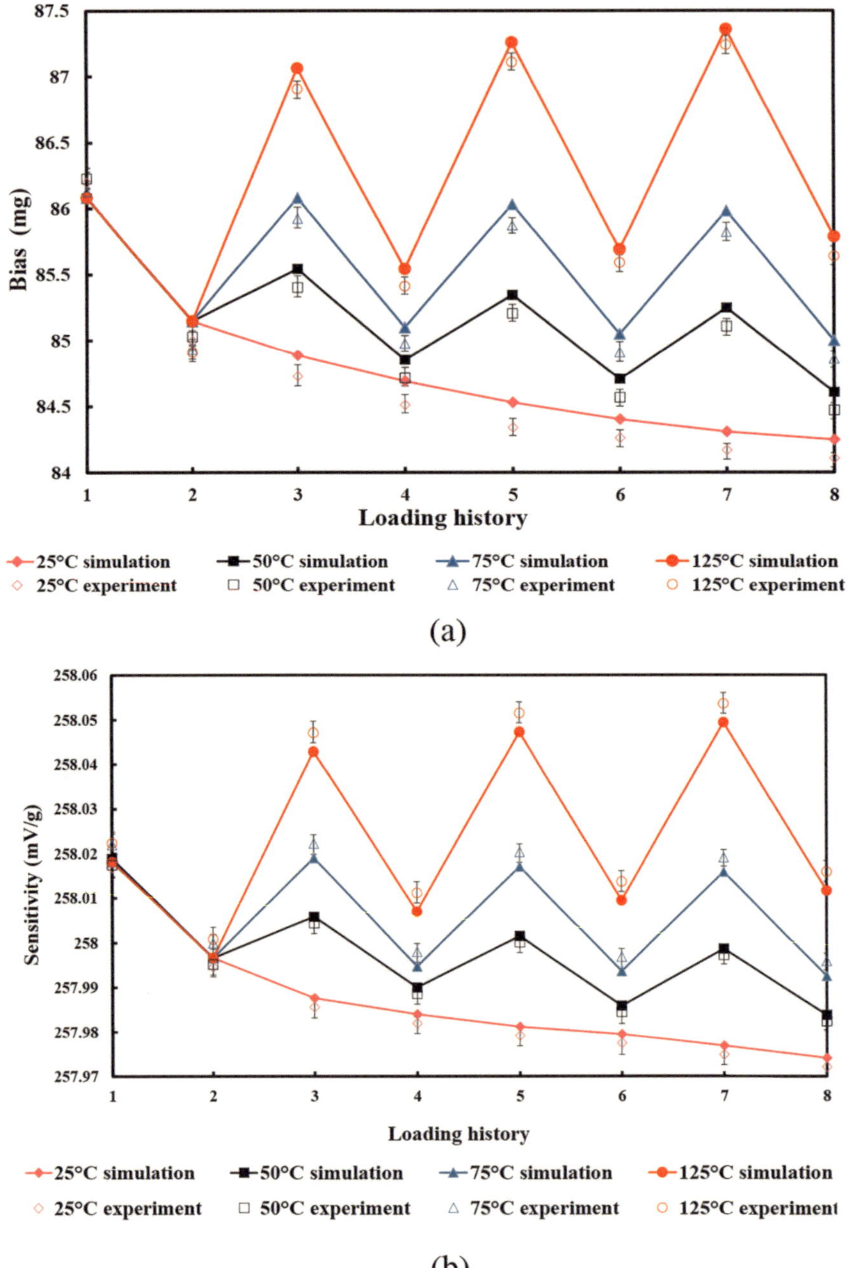

Figure 8.
Results comparison: (a) bias and (b) sensitivity.

geometric dimension, such as gap, width, and length, and consequently induce the performance drift of MEMS device. However, the performance drift induced by CTE is generally very small compared to that induced by CTE mismatch.

Besides the single crystal silicon, there often exist the layers made of other material in a MEMS die, such as glass, SiO_2, metal, and so on. The CTE of these materials is generally different from the single crystal silicon. Even for the borosilicate glass that has a CTE very close to the single crystal silicon, there still exists a CTE mismatch, as shown in **Figure 10**. In literature [57], research shows that the CTE difference between single crystal silicon and borosilicate glass induces bias

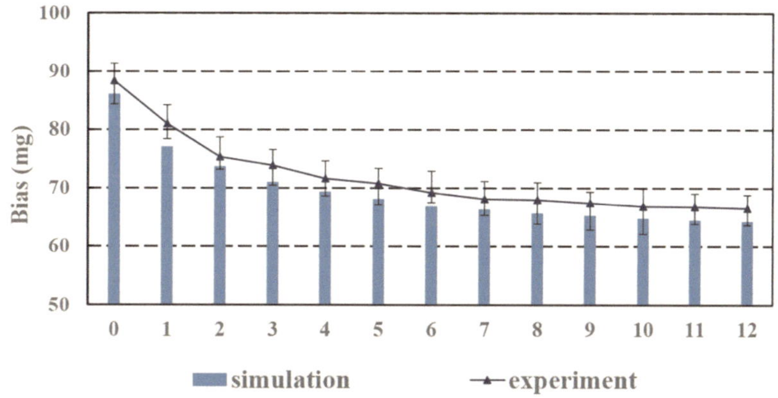

Figure 9.
The variation of the bias of the sensor in 12 months.

TCE	p-type (4 Ω cm, B)	n-type (0.05 Ω cm, P)	p-type (4 Ω cm, B)	n-type (0.05 Ω cm, P)
	First-order ($\times 10^{-6}$/K)		Second-order ($\times 10^{-6}/K^2$)	
TCE_{C11}	−73.25 ± 0.49	−74.87 ± 0.99	−49.26 ± 4.8	−45.14 ± 1.4
TCE_{C12}	−91.59 ± 1.5	−99.46 ± 3.5	−32.70 ± 10.1	−20.59 ± 11.0
TCE_{C44}	−60.14 ± 0.20	−57.96 ± 0.17	−51.28 ± 1.9	−53.95 ± 1.8

Table 2.
Temperature coefficients of the elastic constants given by Bourgeois et al. [56].

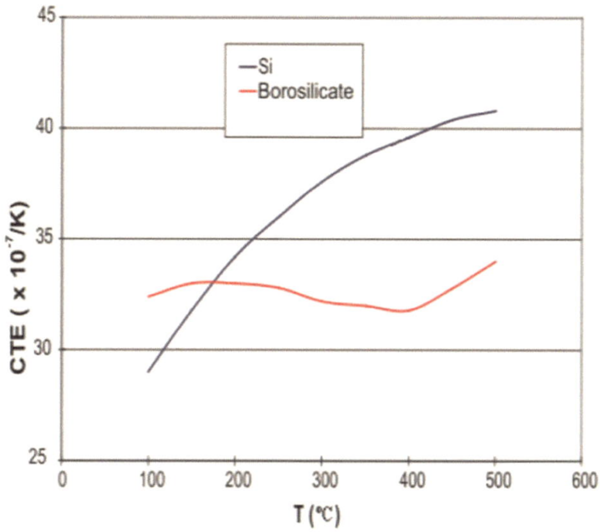

Figure 10.
CTE of silicon and borosilicate glass [61].

drift of MEMS accelerometer. Besides the CTE mismatch in MEMS die, another source of thermal stress/deformation is the CTE disagreement between the MEMS die and the package. The package material is in most cases ceramic, metal, and polymer, whose CTE values differ from the single crystal silicon. For instance, the CTE of a ceramic package is over twice as much as that of single crystal silicon [58]. In order to calculate the thermal stress/deformation, finite element method is

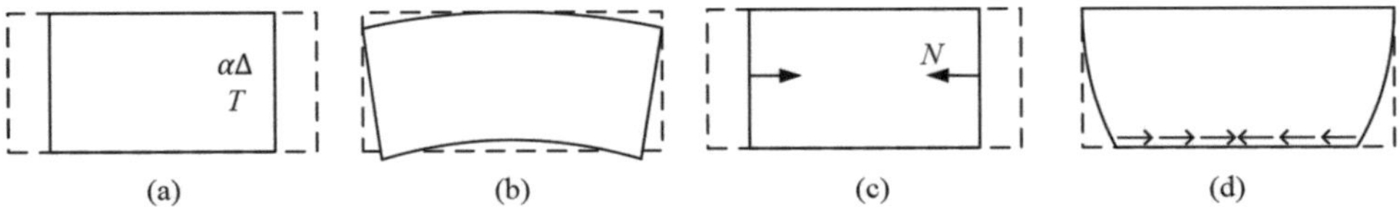

Figure 11.
Four components of the deformation in the die or substrate. (a) Longitudinal normal deformation induced by the thermal expansion, (b) longitudinal normal deformation induced by the shear stress, (c) transverse bending deformation, and (d) longitudinal shearing deformation.

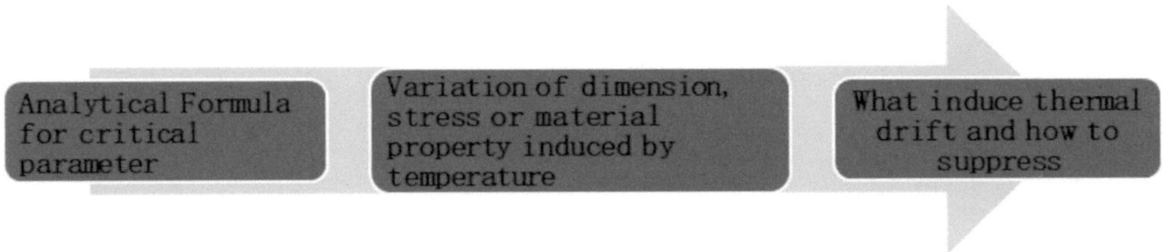

Figure 12.
Thermal drift procedure estimation.

widely employed. However, the finite element method generally generates a model with high degrees and is time-consuming. For the thermal stress/deformation induced by the package, the analytical model based on strength of material is also largely employed, while taking up less time. In the analytical model, the elastic foundation for the adhesive layer is generally employed [59], and the deformation inside the die or substrate is by and large divided into four components, which are shown in **Figure 10**. The four components can be described by the first-order or second-order beam theory [60] (**Figure 11**).

2.2.3 Means of estimating thermal drift

In this section, the procedure estimating the thermal drift is discussed, as shown in **Figure 12**. A case study about the thermal drift of a MEMS capacitive accelerometer is also presented.

The procedure estimating the thermal drift can be divided into three distinct steps:

1. Deriving analytical formulae for critical parameter(s)

This step forms the base of the latter two steps. Defining the precise critical parameter that needs to be derivated depends on application requirement. For instance, the drift of frequency is critical for the application of the MEMS resonator, so the analytical formula for frequency needs to be derived. The imperfection is important in obtaining analytical formulae for critical parameters, especially for the MEMS sensors employing the differential detecting principle. If the imperfection is not considered, such as the asymmetry induced by fabrication error, the bias of MEMS sensors nulls. As such, no result on the thermal drift of bias may be obtained.

2. Calculation of the variation of dimension, stress, or material property induced by temperature

It is critical to calculate the variation of dimension, stress, or material property induced by temperature for estimating the thermal drift. For the material property, its variation with temperature is induced by the temperature coefficient. However, for the dimension and stress, the variation generally needs to be calculated by finite element analysis or other analytical methods. The imperfection is also important for the MEMS sensors employing the differential detecting principle. If the imperfection is not considered, the impact of the variation of dimension and stress is operating in common mode. As such, these variations cannot induce the variation of the bias for the MEMS sensors employing the differential detecting principle.

3. Discussion on the factors affecting the thermal drift and how to suppress the thermal drift

Based on the variation of dimension, stress, or material property induced by temperature, the thermal drift can be acquired by deriving the differential of the temperature. Then, the factors affecting thermal drift and how to suppress this will be discussed.

2.2.4 Case study

In the following, a MEMS capacitive accelerometer is employed to showcase the procedure estimating the thermal drift. The detection of the accelerometer is based on the open-loop differential capacitive principles, as shown in **Figure 13**. The acceleration force makes the proof mass move and is balanced by elastic force generated by folded beams. The moving proof mass changes the capacitances of the accelerometer. The detected variation amplitude of the capacitance difference between capacitors C_A and C_B via modulation and demodulation with preload AC voltage was used to generate the output voltage:

$$V_{out} = G\frac{C_A - C_B}{C_A + C_B} \tag{4}$$

where G is the gain that depends on the circuit parameters.

Based on the detecting principle and the dimension shown in **Figure 13b**, the bias and scale factor are expressed as

$$B = -\frac{K_T e}{m} \tag{5}$$

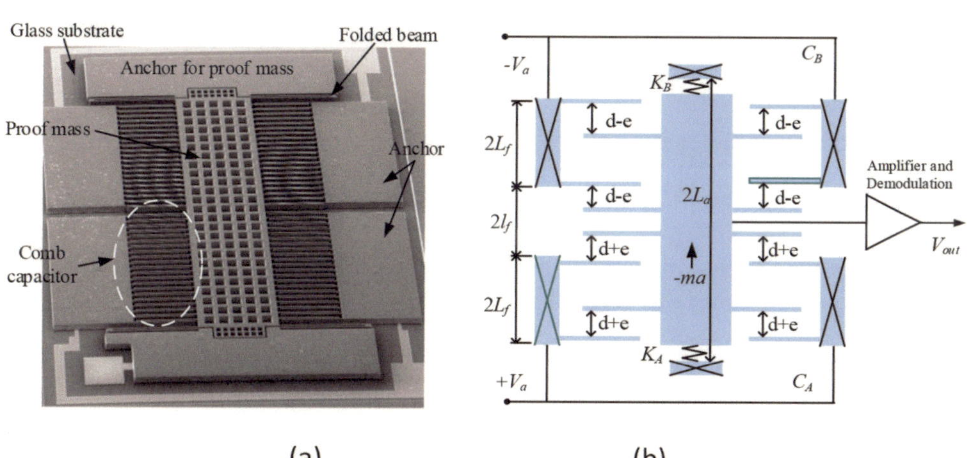

Figure 13.
MEMS capacitive accelerometer employed as case study. (a) SEM picture and (b) open-loop differential capacitive principle. ±Va is preload AC voltage. CA and CB represent capacitors on the bottom and top sides, respectively.

$$k_1 = G\frac{m}{dK_T} \tag{6}$$

where B and k_1 represent the bias and scale factor, respectively; K_T is the total stiffness of the folded beams; m is the total mass of proof mass and moving fingers; e represents the asymmetry of capacitive gap induced by the fabrication error; d is the capacitive gap.

From the equation of bias, it can be seen that if the asymmetry of capacitive gap is not considered, then the bias nulls. In equations of bias and scale factor, the parameters varying with temperature include K_T, e, and d. Variation of the stiffness is induced by TCEM and CTE [7]:

$$TCS = \frac{1}{K_T}\frac{dK_T}{dT} = \alpha_E + \alpha_s \tag{7}$$

The variations of e and d are calculated by analytical method [44]:

$$\Delta e = \frac{K_A - K_B}{K_A + K_B}\left(\alpha_s - \alpha_{eq}\right)\Delta T L_a \tag{8}$$

$$\Delta d = \left(d\alpha_s + \left(l_f + L_f\right)\left(\alpha_{eq} - \alpha_s\right)\right)\Delta T \tag{9}$$

where L_a expresses the distance from the anchor to the midline; L_f denotes the half length of an anchor for fixed fingers; l_f defines the locations of first fixed finger, as shown in **Figure 13b**; α_s indicates the CTE of silicon; α_{eq} is called as equivalent CTE describing the thermal deformation of the top surface of the substrate and calculated by the analytical model for the MEMS die attaching proposed in literature; K_A and K_B stand for the spring stiffness connecting proof mass.

Deriving the differential of the bias to the temperature, the TDB is expressed as

$$\text{TDB} = \frac{\Delta B}{\Delta T} = -\frac{K_T \Delta e}{m \Delta T} = \frac{K_A - K_B}{m}\left(\alpha_{eq} - \alpha_s\right)L_a \tag{10}$$

Deriving the differential of the scale factor to the temperature, the thermal drift of scale factor (TDSF) is expressed as

$$\text{TDSF} = \frac{\Delta k_1}{k_1|_{\Delta T=0}\Delta T} = TCS - \left(\alpha_s + \frac{\left(l_f + L_f\right)\left(\alpha_{eq} - \alpha_s\right)}{d}\right) \tag{11}$$

Due to the asymmetry induced by the fabrication error, K_A and K_B are different from each other, and TDB is proportional to the difference between K_A and K_B. Therefore, the consideration on the imperfection is very important for discussing the thermal drift.

Based on the discussion on TDB and TDSF, the factors affecting thermal drift and method suppressing the thermal drift can be obtained:

1. The model shows that TDB is only caused by thermal deformation, while TDSF consists of two parts caused by stiffness temperature dependence and thermal deformation, respectively. The two parts of TDSF are positive and negative, respectively. However, the second part has a greater absolute value.

2. The first part of TDSF can be reduced by high doping. TDB and the second part of TDSF can be both reduced by soft adhesive die attaching or increasing substrate thickness.

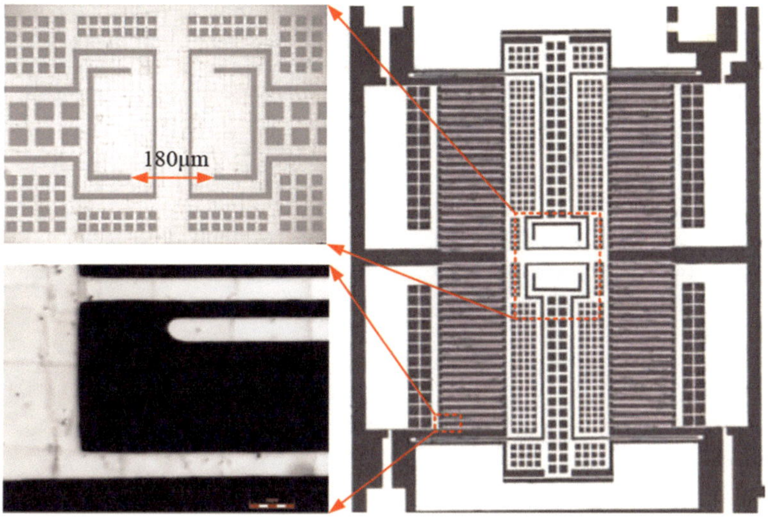

Figure 14.
Accelerometer with optimization for TDSF.

3. In silicon structure, TDB can be reduced by middle-locating anchors for moving electrodes in sensitive direction or decreasing the stiffness asymmetry of springs, while the second part of TDSF can be reduced by middle-locating anchors for fixed electrodes in sensitive direction.

The TDSF of the MEMS capacitive accelerometer can both be induced by the TCEM and the thermal deformation, so the structure of the accelerometer is optimized to make the TCEM and thermal deformation compensate each other, as shown in **Figure 14** [62]. As such, TDSF is suppressed significantly.

3. Conclusions

MEMS devices are an integrated system involving aspects of mechanics, electronics, materials, physics, and chemistry while interacting with the environment. Therefore, their reliability exhibits a great diversity of modes and mechanisms. This is a field open for further research, as it covers a vast area. However, one practical way is to conclude a few failure phenomena of a specific device for providing a guideline to study the similar behaviors appearing in other devices. This chapter only focused on the reliability problem occurring in the micro-accelerometers in storage and the thermal environment. These two factors pose the significant importance on the development of high-end microsensors for high-precise applications. The long-term stability induced by the viscoelasticity of packaging materials was first mentioned in this work to explain the performance shift after a long period of storage. The thermal effects formed by temperature change and structural layout were studied in depth and showed that the drift over temperature may be eliminated by a well-designed structure rather than perfect materials with zero CTE. It is the authors' view that this area should be further researched so as to bridge the diversity of micro-devices and develop standards.

Acknowledgements

This work was supported in part by the National Natural Science Foundation of China (Nos. 51505068 and U1530132).

Conflict of interest

The authors declare no conflict of interest.

Thanks

The authors would like to thank Prof. Xiaoping He, Prof. Lianming Du, and Prof. Zhigui Shi for their assistance in designing the circuit, setting up the experiment, and fabricating the sensors.

Author details

Wu Zhou[1]*, Jiangbo He[2], Peng Peng[3], Lili Chen[3] and Kaicong Cao[1]

1 School of Mechanical and Electrical Engineering, University of Electronic Science and Technology of China, Chengdu, China

2 School of Mechanical Engineering, Xihua University, Chengdu, China

3 School of Mechanical Engineering, Chengdu Technological University, Chengdu, China

*Address all correspondence to: zhouwu916@uestc.edu.cn

References

[1] Müller-Fiedler R, Wagner U, Bernhard W. Reliability of MEMS—A methodical approach. Microelectronics and Reliability. 2002;**42**(9–11): 1771-1776. DOI: 10.1016/S0026-2714(02)00229-9

[2] Van Spengen WM. MEMS reliability from a failure mechanisms perspective. Microelectronics and Reliability. 2003; **43**(7):1049-1060. DOI: 10.1016/S0026-2714(03)00119-7

[3] Van Spengen WM, Puers R, De Wolf I. The prediction of stiction failures in MEMS. IEEE Transactions on Device and Materials Reliability. 2003;**3**(4): 167-172. DOI: 10.1109/TDMR.2003.820295

[4] Tanner DM. MEMS reliability: Where are we now? Microelectronics and Reliability. 2009;**49**(9–11):937-940. DOI: 10.1016/j.microrel.2009.06.014

[5] Fonseca DJ, Sequera M. On MEMS reliability and failure mechanisms. International Journal of Quality, Statistics, and Reliability. 2011;**2011**:7. Article ID 820243. DOI: 10.1155/2011/820243

[6] Huang Y, Vasan ASS, Doraiswami R, Osterman M, Pecht M. MEMS reliability review. IEEE Transactions on Device and Materials Reliability. 2012;**12**(2): 482-493. DOI: 10.1109/TDMR.2012.2191291

[7] He J, Xie J, He X, Du L, Zhou W. Analytical study and compensation for temperature drifts of a bulk silicon MEMS capacitive accelerometer. Sensors and Actuators, A: Physical. 2016;**239**:174-184. DOI: 10.1016/j.sna.2016.01.026

[8] Peng P, Zhou W, Yu H, Peng B, Qu H, He X. Investigation of the thermal drift of MEMS capacitive accelerometers induced by the overflow of die attachment adhesive. IEEE Transactions on Components, Packaging and Manufacturing Technology. 2016;**6**(5):822-830. DOI: 10.1109/TCPMT.2016.2521934

[9] Kahn H, Heuer AH, Jacobs SJ. Materials issues in MEMS. Materials Today. 1999;**2**(2):3-7. DOI: 10.1016/S1369-7021(99)80002-9

[10] Van Driel WD, Yang DG, Yuan CA, Van Kleef M, Zhang GQ. Mechanical reliability challenges for MEMS packages: Capping. Microelectronics and Reliability. 2007;**47**(9–11): 1823-1826. DOI: 10.1016/j.microrel.2007.07.033

[11] Sundaram S, Tormen M, Timotijevic B, Lockhart R, Overstolz T, Stanley RP, et al. Vibration and shock reliability of MEMS: Modeling and experimental validation. Journal of Micromechanics and Microengineering. 2011;**21**(4):045022. DOI: 10.1088/0960-1317/21/4/045022

[12] Tilmans HAC, De Coster J, Helin P, Cherman V, Jourdain A, Demoor P, et al. MEMS packaging and reliability: An undividable couple. Microelectronics and Reliability. 2012;**52**(9–10): 2228-2234. DOI: 10.1016/j.microrel.2012.06.029

[13] Zhang W-M, Yan H, Peng Z-K, Meng G. Electrostatic pull-in instability in MEMS/NEMS: A review. Sensors and Actuators, A: Physical. 2014;**214**: 187-218. DOI: 10.1016/j.sna.2014.04.025

[14] Li J, Broas M, Makkonen J, Mattila TT. Shock impact reliability and failure analysis of a three-axis MEMS gyroscope. Journal of Microelectromechanical Systems. 2014;**23**(2):347-355. DOI: 10.1109/JMEMS.2013.2273802

[15] DelRio FW, Cook RF, Boyce BL. Fracture strength of micro- and

nano-scale silicon components. Applied Physics Reviews. 2015;**2**(2):021303. DOI: 10.1063/1.4919540

[16] Yu L-X, Qin L, Bao A-D. Reliability prediction for MEMS accelerometer under random vibration testing. Journal of Applied Science and Engineering. 2015;**18**(1):41-46. DOI: 10.6180/jase.2015.18.1.06

[17] Wang J, Zeng S, Silberschmidt VV, Guo J. Multiphysics modeling approach for micro electro-thermo-mechanical actuator: Failure mechanisms coupled analysis. Microelectronics and Reliability. 2015;**55**(5):771-782. DOI: 10.1016/j.microrel.2015.02.012

[18] Iannacci J. Reliability of MEMS: A perspective on failure mechanisms, improvement solutions and best practices at development level. Displays. 2015;**37**:62-71. DOI: 10.1016/j.displa.2014.08.003

[19] Shoaib M, Hamid NH, Malik A, Ali NBZ. A review on key issues and challenges in devices level MEMS testing. Journal of Sensors. 2016;**2016**: 1-14. DOI: 10.1155/2016/1639805

[20] Tavassolian N, Koutsoureli M, Papaioannou G, Papapolymerou J. Optimization of dielectric material stoichiometry for high-reliability capacitive MEMS switches. IEEE Microwave and Wireless Components Letters. 2016;**26**(3):174-176. DOI: 10.1109/LMWC.2016.2524596

[21] Cheng YT, Lin L. MEMS Packaging and Thermal Issues in Reliability. Berlin Heidelberg: Springer. 2004. p. 1112. DOI: 10.1007/3-540-29838-X_37

[22] Hsu TR. Reliability in MEMS packaging. In: Proceedings of the IEEE International Conference on Reliability Physics (ICRP'06); 26–30 March 2006; San Jose, CA: IEEE; 2006. pp. 398-402

[23] Ther JBN, Larsen A, Liverød B, et al. Measurement of package-induced stress and thermal zero shift in transfer molded silicon piezoresistive pressure sensors. Journal of Micromechanics and Microengineering. 1998;**8**(2):168-171. DOI: 10.1088/0960-1317/8/2/032

[24] Walwadkar SS, Cho J. Evaluation of die stress in MEMS packaging: Experimental and theoretical approaches. IEEE Transactions on Components and Packaging Technologies. 2006;**29**(4):735-742. DOI: 10.1109/TCAPT.2006.885931

[25] Chuang CH, Lee SL. The influence of adhesive materials on chip-on-board packing of MEMS microphone. Microsystem Technologies. 2012;**18**(11): 1931-1940. DOI: 10.1007/s00542-012-1575-0

[26] Peng P, Zhou W, Yu H, et al. Investigation of the thermal drift of MEMS capacitive accelerometers induced by the overflow of die attachment adhesive. IEEE Transactions on Components, Packaging and Manufacturing Technology. 2017;**6**(5): 822-830. DOI: 10.1109/TCPMT.2016.2521934

[27] Obaid N, Kortschot MT, Sain M. Understanding the stress relaxation behavior of polymers reinforced with short elastic fibers. Materials. 2017; **10**(5):472-486. DOI: 10.3390/ma10050472

[28] Xu C, Segovia J, Kim HJ, et al. Temperature-stable piezoelectric MEMS resonators using integrated ovens and simple resistive feedback circuits. Journal of Microelectromechanical Systems. 2016;**2016**:1-9. DOI: 10.1109/JMEMS.2016.2626920

[29] Ho GK, Sundaresan K, Pourkamali S, et al. Micromechanical IBARs: Tunable high-Q resonators for temperature-compensated reference oscillators. Journal of

Microelectromechanical Systems. 2010; **19**(3):503-515. DOI: 10.1109/JMEMS.2010.2044866

[30] Hopcroft MA, Agarwal M, Park KK, et al. Temperature compensation of a MEMS resonator using quality factor as a thermometer. In: Proceedings of the 19th IEEE International Conference on Micro Electro Mechanical Systems; 22–26 Jan. 2006; Istanbul, Turkey, Turkey: IEEE; 2006. DOI: 10.1109/MEMSYS.2006.1627776

[31] Salvia JC, Melamud R, Chandorkar SA, et al. Real-time temperature compensation of MEMS oscillators using an integrated micro-oven and a phase-locked loop. Journal of Microelectromechanical Systems. 2010; **19**(1):192-201. DOI: 10.1109/JMEMS.2009.2035932

[32] Kim B, Hopcroft MA, Candler RN, et al. Temperature dependence of quality factor in MEMS resonators. Journal of Microelectromechanical Systems. 2008;**17**(3):755-766. DOI: 10.1109/JMEMS.2008.924253

[33] Hopcroft MA, Kim B, Chandorkar S, et al. Using the temperature dependence of resonator quality factor as a thermometer. Applied Physics Letters. 2007;**91**(1):440. DOI: 10.1063/1.2753758

[34] Kose T, Azgin K, Akin T. Temperature compensation of a capacitive MEMS accelerometer by using a MEMS oscillator. In: IEEE International Symposium on Inertial Sensors & Systems. 2016. DOI: 10.1109/ISISS.2016.7435538

[35] Du J, Guo Y, Lin Y, et al. A real-time temperature compensation algorithm for a force-rebalanced MEMS capacitive accelerometer based on resonant frequency. In: IEEE International Conference on Nano/micro Engineered & Molecular Systems; IEEE; 2017. DOI: 10.1109/NEMS.2017.8017009

[36] Melamud R, Chandorkar SA, Kim B, et al. Temperature-insensitive composite micromechanical resonators. Journal of Microelectromechanical Systems. 2009;**18**(6):1409-1419. DOI: 10.1109/JMEMS.2009.2030074

[37] Liu YC. Temperature-compensated CMOS-MEMS oxide resonators. Journal of Microelectromechanical Systems. 2013;**22**(5):1054-1065. DOI: 10.1109/JMEMS.2013.2263091

[38] Ng EJ, Hong VA, Yang Y, et al. Temperature dependence of the elastic constants of doped silicon. Journal of Microelectromechanical Systems. 2015; **24**(3):730-741. DOI: 10.1109/JMEMS.2014.2347205

[39] Hajjam A, Logan A, Pourkamali S. Doping-induced temperature compensation of thermally actuated high-frequency silicon micromechanical resonators. Journal of Microelectromechanical Systems. 2012; **21**(3):681-687. DOI: 10.1109/jmems.2012.2185217

[40] Samarao AK, Ayazi F. Temperature compensation of silicon resonators via degenerate doping. IEEE Transactions on Electron Devices. 2012;**59**(1):87-93. DOI: 10.1109/ted.2011.2172613

[41] Lee H, Kim B, Melamud R, et al. Influence of the temperature dependent nonlinearities on the performance of micromechanical resonators. Applied Physics Letters. 2011;**99**(19):194102. DOI: 10.1063/1.3660235

[42] Zwahlen P, Nguyen A, Dong Y, et al. Navigation grade MEMS accelerometer. IEEE International Conference on MicroElectro Mechanical Systems; Hong Kong, China; 2010. pp. 631-634. DOI: 10.1109/MEMSYS.2010.5442327

[43] Wang J, Li X. Package-friendly piezoresistive pressure sensors with on-chip integrated

packaging-stress-suppressed suspension (PS3) technology. Journal of Micromechanics and Microengineering. 2013;**23**(4):045027. DOI: 10.1088/0960-1317/23/4/045027

[44] Hsieh HS, Chang HC, Hu CF, et al. A novel stress isolation guard-ring design for the improvement of a three-axis piezoresistive accelerometer. Journal of Micromechanics and Microengineering. 2011;**21**(10):105006. DOI: 10.1088/0960-1317/21/10/105006

[45] Hsu WT, Nguyen TC. Stiffness compensated temperature in-sensitive micromechanical resonators. In: IEEE International Conference on Micro Electro Mechanical Systems; Las Vegas, USA; 2002. pp. 731-734. DOI: 10.1109/MEMSYS.2002.984374

[46] Myers DR, Azevedo RG, Chen L, et al. Passive substrate temperature compensation of doubly anchored double-ended tuning forks. Journal of Microelectromechanical Systems. 2012; **21**(6):1321-1328. DOI: 10.1109/JMEMS.2012.2205903

[47] Kim YK, White SR. Stress relaxation behavior of 3501-6 epoxy resin during cure. Polymer Engineering and Science. 2010;**36**(23):2852-2862. DOI: 10.1002/pen.10686

[48] Hu M, Xia Y, Daeffler CS, et al. The linear rheological responses of wedge-type polymers. Journal of Polymer Science Part B: Polymer Physics. 2015; **53**:899-906. DOI: 10.1002/polb.23716

[49] Zhou W, Peng P, Yu H, et al. Material viscoelasticity-induced drift of micro-accelerometers. Materials. 2017; **10**(9):1077-1087. DOI: 10.3390/ma10091077

[50] Zhang X, Park S, Judy MW. Accurate assessment of packaging stress effects on MEMS sensors by measurement and sensor-package interaction simulations. Journal of Microelectromechanical Systems. 2007; **16**(3):639-649. DOI: 10.1109/JMEMS.2007.897088

[51] Krondorfer RH, Kim YK. Packaging effect on MEMS pressure sensor performance. IEEE Transactions on Components and Packaging Technologies. 2007;**30**(2):285-293. DOI: 10.1109/TCAPT.2007.898360

[52] Park S, Liu D, Kim Y, et al. Stress evolution in an encapsulated MEMS package due to viscoelasticity of packaging materials. In: Proceedings of the IEEE Electronic Components and Technology Conference (ECTC'12); 29 May–1 June 2012; San Jose, CA: IEEE; 2012. pp. 70-75

[53] Kim Y, Liu D, Lee H, Liu R, et al. Investigation of stress in MEMS sensor device due to hygroscopic and viscoelastic behavior of molding compound. IEEE Transactions on Components, Packaging and Manufacturing Technology. 2015;**5**(7): 945-955. DOI: 10.1109/tcpmt.2015.2442751

[54] Hopcroft MA, Nix WD, Kenny TW. What is the Young's modulus of silicon? Journal of Microelectro-mechanical Systems. 2010;**19**(2):229-238. DOI: 10.1109/JMEMS.2009.2039697

[55] Dong Y, Zwahlen P, Nguyen AM, et al. Ultra-high precision MEMS accelerometer. In: Proceedings of 2011 16th International Solid-State Sensors, Actuators and Microsystems Conference; 5–9 June 2011; Beijing, China: IEEE; 2011. DOI: 10.1109/TRANSDUCERS.2011.5969218

[56] Bourgeois C, Steinsland E, Blanc N, et al. Design of resonators for the determination of the temperature coefficients of elastic constants of monocrystalline silicon. IEEE International Frequency Control Symposium; IEEE; 1997. DOI: 10.1109/FREQ.1997.639192

[57] Dai G, Li M, He X, et al. Thermal drift analysis using a multiphysics model of bulk silicon MEMS capacitive accelerometer. Sensors and Actuators, A: Physical. 2011;**172**(2):369-378. DOI: 10.1016/j.sna.2011.09.016

[58] Erba A, Maul J, Demichelis R, et al. Assessing thermochemical properties of materials through ab *initio* quantum-mechanical methods: The case of α-Al_2O_3. Physical Chemistry Chemical Physics. 2015;**17**(17):11670-11677. DOI: 10.1039/C5CP01537E

[59] Wang J, Zeng S. Thermal stresses analysis in adhesive/solder bonded bimaterial assemblies. Journal of Applied Physics. 2008;**104**(11):113508. DOI: 10.1063/1.3021357

[60] He J, Zhou W, He X, et al. Analytical model for adhesive die-attaching subjected to thermal loads using second-order beam theory. International Journal of Adhesion and Adhesives. 2018;**82**:282-289. DOI: 10.1016/j.ijadhadh.2018.01.016

[61] SD-2—Glass Substrate for Silicon Sensors. Available from: http://www.hoyaoptics.com/pdf/silicon_sensor.pdf

[62] He J, Zhou W, Yu H, et al. Structural designing of a MEMS capacitive accelerometer for low temperature coefficient and high linearity. Sensors. 2018;**18**(2):643. DOI: 10.3390/s18020643

Section 2

Reliability and Industrial Networks

Chapter 6

A Survivable and Reliable Network Topological Design Model

Franco Robledo, Pablo Romero, Pablo Sartor, Luis Stabile and Omar Viera

Abstract

This work is focused on the resolution of a mixed model for the design of large-sized networks. An algorithm is introduced, whose initial outcomes are promising in terms of topological robustness regarding connectivity and reliability. The algorithm combines the network survivability and the network reliability approaches. The problem of the topological design has been modeled based on the generalized Steiner problem with node-connectivity constraints (GSPNC), which is NP-hard. The aim of this study is to heuristically solve the GSP-NC model by designing low-cost highly connected topologies and to measure the reliability of such solutions with respect to a certain prefixed lower threshold. This research introduces a greedy randomized algorithm for the construction of feasible solutions for the GSP-NC and a local search algorithm based on the variable neighborhood search (VNS) method, customized for the GSP-NC. In order to compute the built network reliabilities, this work adapts the recursive variance reduction (RVR) technique, as a simulation method since the exact evaluation of this measurement is also NP-hard. The experimental tests were performed over a wide set of testing cases, which contained heterogeneous topologies, including instances of more than 200 nodes. The computational results showed highly competitive execution times, achieving minimal local optimal solutions of good quality fulfilling the imposed survivability and reliability conditions.

Keywords: survivability, meta-heuristic, VNS, VND, reliability, simulation, RVR

1. Introduction

The arrival of the optical fiber allowed for an enormous increase on communication line bandwidth. This naturally led to deploying networks with dispersed topologies, as scarce (even unique) paths linking the diverse sites are enough to fulfill all the requirements regarding data exchanges. Yet dispersed topologies have a problem: the ability of the network to keep all sites connected is affected by the failure of few (even one single) communication links or switch sites. Before the introduction of the optical fiber, topologies were denser; upon failure of few links, there was a probable reduction of throughput, yet keeping all sites connected. With disperse fiber-based designs, the network behaves much more as an "all-or-nothing" service; either, it fulfills all requirements of connectivity and bandwidth, or it fails to connect some sites. Therefore, the problem of designing networks with minimal costs and reliability thresholds has since gained relevance.

In view of the above, the networks must continue to be operative even when a component (link or central office) fails. In this context, survivability means that a certain number of pre-established disjoint paths among any pair of central offices must exist. In this case, node-disjoint paths will be required, which show a stronger constraint than the edge-disjunction ones. Assuming that both the links and the nodes have associated certain operation probabilities (elementary reliability), the main objective is to build a minimum-cost sub-network that satisfies the node-connectivity requirements. Moreover, its reliability, i.e., the probability that all sites are able to exchange data at a given point in time, ought to surpass a certain lower bound pre-defined by the network engineer. In this way the model takes into account the robustness of the topology to be designed by acknowledging its structure even in probabilistic terms.

1.1 Aim and objectives

The aim of this chapter is to introduce an algorithm that combines the two approaches so as to achieve the design of robust networks and test it on a number of instances that are representative of communication network problems. On the one hand, the network must be highly reliable from a probabilistic point of view (its reliability) assuming that the probabilities of failure of all links and sites are known. On the other hand, the network structure must be topologically robust; for this, node or edge connectivity levels between pairs of distinguished nodes are required. This means that between all pairs of distinguished nodes there exists a given number of edge paths or disjoint nodes. Then, once a minimum threshold for reliability (e.g., 0.98) is set, the algorithm here introduced:

i. Constructs feasible low-cost solutions that satisfy the connectivity levels (disjoint paths) between pairs of distinguished nodes of the network (terminal nodes).

ii. The reliability of the network built meets the pre-established threshold, thus achieving fault-resistant networks according to both approaches.

Performed research indicates that literature pertaining to algorithms on design topologies which consider both approaches (survivable networks and network reliability) is scarce. The works on the design of robust networks in general fix a level of node/edge global connectivity of the network and try to design a network at the lowest possible cost that satisfies that level (e.g., 2-node-connectivity) [1]. Nevertheless, there are contexts where this combination of approaches is imperative and demanded. For example, in the context of military telecommunication networks, it is required that the networks are topologically very robust (e.g., 3-node-connectivity) and at the same time that they are extremely reliable from the network reliability approach's point of view, surpassing very high reliability levels. Another example of the application of the combined model is the logical distribution of highly dangerous merchandise on a country's roads. In such a context, two things are desirable: high reliability in the connection of points of distribution (i.e., "reliable" roads) and high levels of connectivity between the points that must exchange cargo (availability of alternative roads to possible road cuts, traffic saturation, etc.).

1.2 Problem definition

Formally, the proposed model that combines the network survivability and network reliability approaches is the following:

Consider:

- $G = (V,E)$ as a nondirected simple graph.
- $T \subseteq V$ as a subset of distinguished nodes (denominated terminals).
- $C = \{c_{ij}\}_{(i,j)\in E}$ as a nonnegative cost matrix associated with the edges of G.
- $R = \{r_{ij}\}_{i,j\in T}$ as a node-connectivity requirements matrix among pairs of terminal nodes in which at least r_{ij} node-disjoint paths are required to be communicating bearing i and j in the solution.

In addition to the above, suppose that the edges of E and the nodes of $V\backslash T$ (usually called Steiner nodes) have associated operation probabilities given by the vectors: $P_E = \{p_e\}_{e\in E}$ and $P_{V\backslash T} = \{p_v\}_{v\in V\backslash T}$, respectively, where the failures are assumed to be statistically independent. Given a certain probability p_{min} set as reliability lower threshold, the objective is to find a subgraph $G_S \subseteq G$ of minimum cost that satisfies the node-connectivity requirement matrix R and furthermore its T-terminal reliability. The latter is defined as the probability that the partial graph $G_R \subseteq G_S$, obtained by randomly dropping edges and nodes from G_S with probabilities given by $1-p_e$ and $1-p_v$, respectively, connects all nodes in T [2]. The T-terminal reliability $R_T(G_S)$ has to satisfy $R_T(G_S) \geq p_{min}$ (i.e., the probability that all nodes in T are connected by working edges exceed p_{min}). This model will be referred to as "generalized Steiner problem with survivable and reliable constraints," GSP-SRC.

1.3 Chapter organization

The remainder of this chapter is structured as follows:

- Sections 2 and 3 present a background of the meta-heuristic algorithm applied and the means to compute the network reliability, respectively.
- Section 4 introduces the algorithm required to solve the GSP-SRC.
- Section 5 deals with the experimental results obtained over a set of heterogeneous test instances as well as the most important contributions and conclusions of this work.

2. Greedy randomized adaptive search procedure (GRASP)

Greedy randomized adaptive search procedure (GRASP) is a well-known metaheuristic, i.e., a particular method to find sufficiently good solutions to optimization problems, that has been successfully used to solve many difficult combinatorial optimization problems. It is an iterative multi-start process which operates in two phases, namely, the construction and the local search phases.

In the construction phase, a feasible solution is built whose neighborhood is then explored in the local search phase [3]. A neighborhood of a certain solution S is a set of solutions that differ from S in well-defined forms (e.g., replacing any link by a different one, replacing "stars" by "triangles," and so on). Regarding optimization, different neighborhoods of S will not, in general, share the same local minimum. Thus, local optima trap problems may be overcome by deterministically changing the neighborhoods [4, 5].

Figure 1 illustrates a generic GRASP implementation pseudo-code. The GRASP takes as input parameters the following:

- The candidate list size.
- The maximum number of GRASP iterations.
- The seed for the random number generator. After reading the instance data (line 1), the GRASP iterations are carried out in lines 2–6. Each GRASP iteration consists of the construction phase (line 3), the local search phase (line 4), and, if necessary, the incumbent solution update (lines 5 and 6).

In the construction phase, a feasible solution is built, with one element at a time. At each step of the construction phase, a candidate list is determined by ordering all non-already selected elements with respect to a greedy function that measures the benefit of including them in the solution. The heuristic is adaptive because the benefits associated with every element are updated at each step to reflect the changes brought on by the selection of the previous elements. Then, one element is randomly chosen from the best candidate list and added into the solution. This is the probabilistic component of GRASP, which allows for different solutions to be obtained at each GRASP iteration but does not necessarily jeopardize the power of the adaptive greedy component.

The solutions generated by the construction phase are not guaranteed to be locally optimal with respect to simple neighborhood definitions. Hence, it is beneficial to apply a local search to attempt to improve each constructed solution. A local search algorithm works in an iterative fashion by successively replacing the current solution by a better solution from its neighborhood. It terminates when there is no better solution found in the neighborhood. The local search algorithm depends on the suitable choice of a neighborhood structure, efficient neighborhood search techniques, and the starting solution.

The construction phase plays an important role with respect to this last point, since it produces good starting solutions for local search. Normally, a local optimization procedure, such as a two-exchange, is employed. While such procedures can require exponential time from an arbitrary starting point, experience has shown that their efficiency significantly improves as the initial solutions improve. Through the use of customized data structures and careful implementation, an efficient construction phase that produces good initial solutions for efficient local search can be created. The result is that often many GRASP solutions are generated in the same amount of time required for the local optimization procedure to converge from a single random start. Furthermore, the best of these GRASP solutions is generally significantly better than the solution obtained from a random starting point.

```
Procedure GRASP(ListSize,MaxIter,RandomSeed) ;
1: InputInstance();
2: for k = 1 to MaxIter do
3: InitialSolution = ConstructGreedyRandomizedSolution(ListSize,RandomSize);
4: LocalSearchSolution = LocalSearch(InitialSolution);
5: if cost(LocalSearchSolution) < cost(BestSolutionFound) then
6: UpdateSolution(BestSolutionFound,LocalSearchSolution);
7: end for;
8: return BestSolutionFound;
```

Figure 1.
GRASP pseudo-code.

3. Recursive variance reduction technique

Recursive variance reduction (RVR) is a Monte Carlo simulation method for network reliability estimation [6]. It has shown excellent performance relative to other estimation methods, particularly when component failures are rare events.

It is a recursive method that works with probability measures conditioned to the operation or failure of specific cut-sets. A cut-set is a set of links (or nodes) such that the failure of any of its members results in a failure state for the overall network. RVR computes the unreliability (i.e., 1—reliability) of a network by finding a cut-set and recursively invoking itself several times, based on exhaustive and mutually exclusive combinations of up-and-down states for the members of the cut-set that cover the "cut-set fail" state space (i.e., a partition of the latter). While finding the cut-set and linking the recursion results introduce some overhead compared to other methods (e.g., crude Monte Carlo), RVR achieves significant reductions of the unreliability estimator variance, particularly in the (realistic) setting where failures are rare events. This allows for the use of smaller sample sizes, eventually beating the alternative methods in the trade-off between processing time and precision.

4. The algorithmic solution for the GSP-SRC

4.1 Network design algorithm

NetworkDesign is the main algorithm which iteratively executes the different phases that solve the GSP-SRC. The algorithm (shown in **Figure 2**) receives as entry *G* the original graph, *MaxIter* the number of iterations that is going to be executed, *k* an integer (parameter of the construction phase), the *threshold* of *T*-terminal reliability required, and the number of replications used in reliability phase.

```
Procedure NetworkDesign;
Input: G, C, k, MaxIter, threshold, simiter;
i ← 0; P ←∅; L_Sol ←∅;
while (i < MaxIter) do
[G*sol,P] ← ConstructionPhase(G,C,k);
Gsol ← SurvivabilityOptimizerPhase(G*sol,P);
R*T(Gsol) ← RealiabilityPhase(Gsol,simiter);
if (R*T(Gsol) ≥ threshold) then L_Sol ← Add(Gsol) ;
else Discard(Gsol);
i ← i + 1;
end while;
return L_Sol;
end NetworkDesign;
```

Figure 2.
Global algorithm.

Each iteration computes:

i. Construction phase

ii. Survivability optimizer phase

iii. Reliability phase

The construction phase takes G as the input and returns a topology satisfying the node-connectivities given by R. Since the solution built by the construction phase is not even a local optimum, in order to improve this solution, survivability optimizer phase searches for a local optimum solution by means of a variable neighborhood search (VNS) [7, 8] algorithm designed specifically for the GSP-NC. Finally, the reliability phase is computed evaluating the T-terminal reliability of the solution achieved in (*ii*). If it surpasses the prefixed threshold, then the local optimal solution is added into the collection *L_Sol*; otherwise, it is discarded. The algorithm returns a list *L_Sol* of feasible solutions that satisfy the pre-established survivability and reliability requirements.

4.2 Construction phase algorithm

The algorithm (shown in **Figure 3**) takes as input the graph G of feasible connections, the matrix of connection costs C, the matrix R of connection requirements between terminal nodes, and a parameter k. The current solution G_{sol} is initialized with the terminal nodes without any connection among them. An auxiliary matrix M is initialized with the values of R. This is used with the purpose of maintaining on each step the connection requirements not yet satisfied between nodes of T. The paths found on each iteration are stored in a data structure P. Iteratively, the construction phase searches for node-disjoint paths between terminal nodes of T that have not yet satisfied their connection requirements. The algorithm chooses on each iteration a pair of such terminal nodes $i,j \in T$. The current solution is updated by adding a new low-cost node-disjoint path between the chosen nodes. For this, an

```
Procedure ConstructionPhase;
Input: G, C, R, k;
Gsol ← (T,∅); M ← R; Pij ←∅∀i,j ∈ T;
while ∃mij > 0 do
Let i,j ∈ T be two nodes that satisfy mij > 0;
G* ← (G \Pij);
C* ← C |G*\Gsol; /* C*uv = 0 if (u,v) ∈ Gsol */
Lp ← the k shortest paths from i to j on G*, considering C*;
p ← SelectRandom(Lp);
Gsol ←Gsol∪ {p}; Pij ←Pij∪ {p};
M ← UpdateMatrix(Gsol,M,R,p);
end while;
return Gsol, P;
end ConstructionPhase
```

Figure 3.
Construction phase.

extension of the Takahashi-Matsuyama algorithm is employed in order to efficiently compute the k shortest node-disjoint paths from i to j (lines 3–9). These paths are stored in a restricted candidate list L_p. A path is randomly selected from L_p and incorporated into G_{sol}. This process is repeated until all the connection requirements have been satisfied; then, the feasible solution G_{sol} and the set of node-disjoint paths $P = \{P_{ij}\}i,j \in T$ are returned.

4.3 Survivability optimizer phase: VNS algorithm for the GSP-SRC

The variable neighborhood search algorithm is sustained on the idea of systematically changing the neighborhood at the moment of performing the local search and requires therefore a finite set of different pre-defined neighborhoods.

VNS is based on three simple facts:

- A local minimum with respect to a neighborhood structure, which is not necessary to be a local minimum with respect to another one.
- A global minimum which is a local minimum with respect to all the possible neighborhood structures.
- In many problems the local minimum with respect to one or several neighborhood structures are relatively close.

In this work, the deterministic variant called VNS descent was used (variable neighborhood descent, VND). It consists of iteratively replacing the current solution with the local search result as long as improvements are verified. If a

```
Procedure VND_SurvivabiliyOptimizerPhase;
Input: Gsol, P, cls;
    cost ← cost(Gsol); kmax ← 2;
    Gsol ← SwapKeyPathLocalSearch(Gsol,P);
    cost ← cost(Gsol); notimprove ← 0;
    while (notimprove < kmax) do
    G*sol ← cls[kmax] → LocalSearch(Gsol,k);
    cost ← cost(Gsol); newcost ← cost(G*sol);
    if (newcost < cost) then
              cost ← newcost; notimprove ← 0;
              Gsol ← G*sol;
    else notimprove ← notimprove + 1;
    k ← (k + 1) mod kmax;
    end while;
    return Gsol;
end VND_SurvivabiliyOptimizerPhase;
```

Figure 4.
Survivability phase.

neighborhood structure change is performed in a deterministic way every time a local minimum is reached, the descent variable neighborhood search is obtained. The final solution given by the VND is a local minimum with respect to all the considered neighborhoods. Next, the VND customization for the GSPNC will be explained. Here, the VND uses three local searches: SwapKeyPathLocalSearch, KeyPathLocalSearch, and KeyTreeLocalSearch. Details on these local searches, their respective neighborhood structures, and the extension of the Takahashi-Matsuyama algorithm mentioned above can be found in [9].

The algorithm (shown in **Figure 4**) receives as input G_{sol} the initial solution graph, P the matrix of paths (both outputs of construction phase), and *cls* the set of local searches. Initially, the cost of G_{sol} is computed, and the local search SwapKeyPathLocalSearch is applied to it. SwapKeyPathLocalSearch uses P the path's matrix as input (lines 1–2). Since only this local search uses the P information, it is executed only at the beginning of the algorithm in a single time. However, its incorporation is fundamental for the purpose of achieving important improvements in the initial solutions generated by the algorithm of construction. Line 3 computes the new cost of G_{sol} and *notimprove* is initialized to zero. In cycles 4–12, the *k*th local search is performed to find a better solution (line 5) until no more improvements can be found by exploring the neighborhood set. If an improvement is achieved, the current cost is updated, *notimprove* is reset to zero, and the new solution G_{sol} is actualized (lines 8–9). In the case of not having improvements, *notimprove* is increased by one (line 10). Finally, regardless of the fact that improvements have been made,

```
Procedure RVR_ReliabilityPhase;
Input: G_sol, N;
1:    seed ← GenerateSeed();
2:    s ← 0; ss ← 0;
3:    for l = 1 to N
4:    G'_sol ←G_sol;
/* RVR computes the anti-reliability */
5:    x ← 1 – RVR(G'_sol,P_E,P_V\T);
6:    s ← s + x;
7:    ss ← ss + (x^2);
8:    end for;
/*  RT(Gsol) estimation */
9:  R*_T( Gsol) ← s/N;
10:  variance ← (1/(N×(N–1))) (ss–(s^2/N));
11:  return R*_T(G_sol), variance;
end RVR_ReliabiliyPhase;
```

Figure 5.
Reliability phase.

the next neighborhood is explored in a circular form (line 11). In this way, and unlike the generic VND, when finding an improvement, the algorithm continues the search for new solutions in the following neighborhood instead of returning to neighborhoods already explored. Once the loop is finalized (lines 4–12), line 13 returns the best solution found by the VND.

4.4 Reliability phase: RVR algorithm for the GSP-SRC

The reliability of G_{sol} is estimated by using the RVR method. The construction of the method is done so as to obtain an estimation of the Q_K measurement (anti-reliability for a K set of terminals). As such, the RVR method is used in order to estimate the T-terminal reliability of G_{sol} assuming that the terminal nodes T are perfect (i.e., do not fail), and the edges as well as \the Steiner nodes (nodes of $V \setminus T$) have operation probabilities $P_E = \{p_e\}e \in E$ and $PV \setminus T = \{p_v\}v \in V \setminus T$, respectively.

Details and properties of this adaptation can be found in [10]. **Figure 5** shows the pseudo-code of the R_T (G_{sol}) estimated computation using the RVR method.

5. Computational results and discussion

To the best of the authors' knowledge, there are no efficient optimization algorithms that design robust networks combining the requirement of network reliability and high levels of topological connectivity between pairs of distinguished nodes (topological design of survivable networks). Critical applications such as military communication networks, transport and distribution of highly risky or critical products/substances, or circuit design with high requirements of redundancy (airplanes), among others, were influential factors that facilitated the combination of efficient network design with high levels of connectivity that exceeded a pre-established threshold of reliability. Given the lack/absence of real cases in the literature and plausible instances to be used as a benchmark for the combined problem GSP-SRC, 20 instances of the traveling salesman problem (TSP) from the TSPLIB library have been selected [11], and for each of these, three GSP-SRC instances have been generated by randomly marking 20, 35, and 50% out of the nodes as terminal. The 20 TSP test instances were chosen so that their topologies are heterogeneous and their numbers of nodes (which range from 48 to 225) cover typical applications of the GSP problem in telecommunication settings. The acronyms used for the 20 instances in the TSPLIB library are att48, berlin52, brazil58, ch150, d198, eil51, gr137, gr202, kroA100, kroA150, kroB100, kroB150, kroB200, lin105, pr152, rat195, st70, tsp225, u159, and rd100. The connectivity requirements were randomly set in $r_{ij} \in \{2,3,4\}$ and $\forall i,j \in T$.

It is worth noting that in all cases the best solutions attained by the VND algorithm were topologically minimal (i.e., feasibility is lost upon removal of any edge). The improvement percentage of the VND algorithm with respect to the solution cost delivered by the construction phase ranged from 25.25 to 39.84%, depending on the topological features of the instance, showing thus the potential of proposed VND in improving the quality of the starting solution.

For the instances in which the average T-terminal reliability of the *L_Sol* solutions set returned by NetworkDesign was computed, it widely surpassed the 85% prefixed threshold. Particularly, in those instances in which the operation probabilities of nodes and edges were set at 99 and 90%, respectively, the average T-terminal reliability was bounded by 86.0 and 96.7%. On the other hand, when setting the values of the operation probabilities of nodes and edges at 99 and 95%, respectively, the average T-terminal reliability was bounded by 99.1 and 99.6%. In all these evaluated cases, the average variance was small, lower than $1.0E{-}05$.

The average times per iteration reached by the NetworkDesign approximated algorithm were below 173 seconds in all test cases.

6. Conclusions

This chapter discussed the generalized Steiner problem with survivable and reliable constraints (GSP-SRC), combining the network survivability and network reliability approaches, to design large-size reliable networks. In addition, a heuristic algorithm was introduced in order to solve the GSP-SRC that combines the GRASP and VNS meta-heuristics (for design and optimization) and the RVR simulation technique for estimating the network reliability of the solutions built.

After testing the algorithm hereby introduced on 60 instances of the GSP-SRC problem, all solutions shared the following desirable facts:

- The solution attained by the VND algorithm was topologically minimal.
- The solution was significantly improved by the VND algorithm with respect to the solution built by the construction phase (25.25–39.84%).
- The prefixed threshold of T-terminal reliability was surpassed in all applicable cases with variances lower than 1.0E − 05.

Given the fact that the GSP-SRC is a NP-hard problem, it is the authors' view that the times per iteration are highly competitive (less than 173 seconds in the worst case).

Author details

Franco Robledo[1], Pablo Romero[1], Pablo Sartor[2*], Luis Stabile[1] and Omar Viera[1]

1 Facultad de Ingeniería, Instituto de Computación, Universidad de la República, Montevideo, Uruguay

2 Departamento de Operaciones, IEEM Business School, Universidad de Montevideo, Montevideo, Uruguay

*Address all correspondence to: psartor@um.edu.uy

References

[1] Mechthild S. Design of Survivable Networks. Lecture Notes in Mathematics. Verlag Berlin, Heidelberg: Springer; 1992;**1531**:1. DOI: 10.1007/BFb0088963. ISBN: 978-3-540-56271-9, ISBN: 978-3-540-47500-2. ISSN: 0075-8434

[2] Ball MO, Colbourn CJ, Provan SJ. Network Reliability. University of Maryland, Systems Research Center; 1992. https://drum.lib.umd.edu/bitstream/handle/1903/5255/TR_92-74.pdf?sequence=1&isAllowed=y

[3] Resende MG, Ribeiro CC. Optimization by GRASP—Greedy Randomized Adaptive Search Procedures, Computational Science and Engineering. New York: Springer-Verlag; 2016

[4] Duarte A, Mladenovic N, Sanchez-Oro J, Todosijevic R. Variable Neighborhood Descent. Cham: Springer International Publishing; 2016

[5] Salhi S. Handbook of metaheuristics (2nd ed.). Journal of the Operational Research Society. 2014;**65**(2):320-320

[6] Cancela H, El-Khadiri M. The recursive variance-reduction simulation algorithm for network reliability evaluation. IEEE Transactions on Reliability. 2003;**52**(2):207-212

[7] Mladenovic N, Hansen P. Variable neighborhood search. Computers & OR. 1997;**24**(11):1097-1100. DOI: 10.1016/S0305-0548(97)00031-2

[8] Hansen P, Mladenović N. Variable neighborhood search. In: Martí R, Pardalos P, Resende M. editors. Handbook of Heuristics; Springer, Cham. 2018. pp. 759-787. DOI: 10.1007/978-3-319-07124-419. Print ISBN: 978-3-319-07123-7. Online ISBN: 978-3-319-07124-4

[9] Robledo F. GRASP heuristics for wide area network design [PhD thesis]. Rennes, France: INRIA/IRISA, Université de Rennes I; 2005

[10] Laborde SSR, Rivoir A. Diseño de Topologías de Red Confiables, tesis f657, INCO, Facultad de Ingeniería, UdelaR, 2006 (Advisors: Robledo F, Viera O)

[11] Reinelt G. n.d. Available from: https://wwwproxy.iwr.uni-heidelberg.de/groups/comopt/software/TSPLIB95/ [Accessed: 25 February 2019]

Chapter 7

Treatment of Uncertainties in Probabilistic Risk Assessment

Vanderley de Vasconcelos, Wellington Antonio Soares, Antônio Carlos Lopes da Costa and Amanda Laureano Raso

Abstract

Probabilistic risk assessment (PRA), sometimes called probabilistic safety analysis, quantifies the risk of undesired events in industrial facilities. However, one of the weaknesses that undermines the credibility and usefulness of this technique is the uncertainty in PRA results. Fault tree analysis (FTA) and event tree analysis (ETA) are the most important PRA techniques for evaluating system reliabilities and likelihoods of accident scenarios. Uncertainties, as incompleteness and imprecision, are present in probabilities of undesired events and failure rate data. Furthermore, both FTA and ETA traditionally assume that events are independent, assumptions that are often unrealistic and introduce uncertainties in data and modeling when using FTA and ETA. This work explores uncertainty handling approaches for analyzing the fault trees and event trees (method of moments) as a way to overcome the challenges of PRA. Applications of the developed frameworks and approaches are explored in illustrative examples, where the probability distributions of the top event of fault trees are obtained through the propagation of uncertainties of the failure probabilities of basic events. The application of the method of moments to propagate uncertainty of log-normal distributions showed good agreement with results available in the literature using different methods.

Keywords: accident, fault tree, event tree, nuclear, probabilistic risk assessment, reliability, uncertainty

1. Introduction

Accidents at industrial facilities may result in serious consequences to workers, public, property, and the environment. Risk management approaches are aimed at insuring that processes and systems are designed and operated to meet "acceptable or tolerable risk levels" as required by regulatory bodies. Risk assessment usually encompasses the following steps: hazard identification, risk analysis, and risk evaluation. When the risk evaluation is carried out in a quantitative way, the risk assessment is considered a probabilistic risk assessment (PRA).

Fault tree analysis (FTA) and event tree analysis (ETA) are the most used techniques in PRAs. However, uncertainties in PRAs may lead to inaccurate risk level estimations and consequently to wrong decisions [1]. Lack of knowledge about systems under study during the PRAs is one of the main causes of uncertainties, which leads to simplification of assumptions, as well as imprecision and inaccuracies in the parameters used as inputs to PRA (e.g., component reliabilities, failure probabilities, and human error rates).

A framework to use the method of moments for determining the likelihoods of different outcomes from event trees in an uncertain data environment using fault trees is described in this work. Illustrative examples using this approach for propagating uncertainty in basic events of fault trees, following log-normal distributions, are also presented. The probability distributions of top events are compared with analyses available in the literature using different approaches, such as Monte Carlo simulation and Wilks and Fenton-Wilkinson methods.

2. Basics of risk assessment

There are many concepts of risk used in different scientific, technological, or organization areas. In a general sense, risk can be defined as the potential of loss (e.g., material, human, or environment) resulting from exposure to a hazard (e.g., fire, explosion, or earthquake). Sometimes, risk is measured through the assessment of the probability of occurrence of an undesired event and the magnitude of consequences [2]. In this way, risk assessment encompasses the answers to the following questions [3]:

- What can go wrong that may lead to an outcome of hazard exposure (scenario S_i)?
- How likely is this to happen, and if so, what is its frequency (F_i)?
- If it happens, what are the likely consequences (C_i)?

Therefore, risk, R_i, for a scenario S_i, can be quantitatively expressed as function of these three variables, as given by Eq. (1):

$$R_i = f(S_i, F_i, C_i). \tag{1}$$

According to Christensen et al. [4], hazard is an inherent property of a risk source potentially causing consequences or effects. This hazard concept does not include the probability of adverse outcome, which is the core difference from risk term. In this chapter, hazard is then considered as the properties of agents or situations capable of having adverse effects on facilities, human health, or environment, such as dangerous substance, sources of energy, or natural phenomena.

2.1 Probabilistic risk assessment (PRA)

PRA provides an efficient way for quantifying the risks, even in an environment of uncertainties regarding possible scenarios, data, or modeling. Risk assessment is part of risk management carried out before deciding about risk treatment and prioritizing actions to reduce risks (risk-based decision-making). **Figure 1** shows a framework for PRA under uncertainty environment [5, 6].

PRA starts with the hazard identification and scenario development, proceeds through quantification of frequencies and consequences, and ends with risk analysis and evaluation [5].

The first step of a PRA process consists of finding, recognizing, and recording risk sources (hazard identification). The accident scenario development (sequence or chain of undesired events) consists of identifying the initiating events (IEs) and

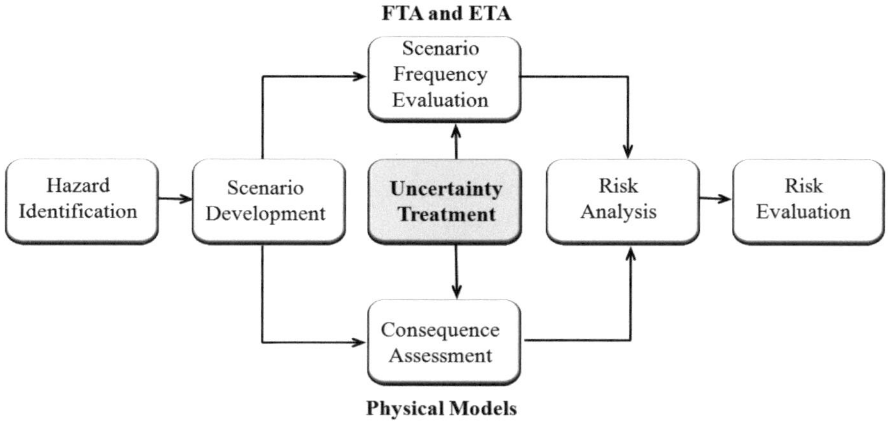

Figure 1.
Framework for probabilistic risk assessment under uncertainty (based on Refs. [5, 6]).

the sequences of events following these IEs. The latter are the critical events that initiate an accident, such as pipe rupture, overpressures, or explosion. The sequences of events are the combinations of success or failure of the barriers or controls requested by IEs (defense-in-depth layers), for example, emergency shutdown systems, human actions, or physical protection. Each sequence can lead to a desired or undesired outcome (end state) such as uncontrollable release of toxic gases, radiation exposure, or facility shutdown [6].

Fault trees (FTs) and event trees (ETs) are often used in PRAs for quantifying the likelihood of event sequences. FTs quantify frequencies or probabilities of top events (such as IEs or failure of defense-in-depth layers) through causal relationship of basic events (e.g., system components, human actions, or subsystems). ETs identify and evaluate each sequence frequency using data generated by FTs [5].

The consequence assessment of each accident scenario to people, property, or environment depends on many factors, such as magnitude of the event, number of people exposed to harm, atmospheric conditions, mitigating measures, etc. The consequence modeling involves the use of analytical or empirical physical or phenomenological models, such as plume dispersion, blast impact (TNT equivalent), or Monte Carlo simulation [7, 8].

Risk analysis is the combination and integration of the probabilities (or frequencies) and the consequences for identified hazards, taking into account the effectiveness of any existing controls and barriers. It provides an input to risk evaluation and decisions about risk treatment and risk management strategies [6].

There are many uncertainties associated with the analysis of risk related to both probability and consequence assessments. An assessment of uncertainties is necessary to perform risk evaluation and to take decisions. The major categories of uncertainties are associated with data, methods, and models used to identify and analyze risks. Uncertainty assessment involves the determination of the variation or imprecision in the results, based on uncertainties of basic parameters and assumptions used in the analyses. Uncertainty propagation of failure probability distributions in FTs and ETs, as well as variability analysis of physical processes (named stochastic uncertainty) and the uncertainties in knowledge of these processes (named epistemic uncertainty), have to be properly accounted for in PRA results [9].

Risk evaluation involves comparing estimated levels of risk with risk criteria defined, once the context of analysis has been established. Uncertainty assessment is important to adjust the categorization of the risk ranking, supporting the decision-makers in meeting risk criteria of standards and guidelines, as well as in visualizing and communicating risks [10].

2.2 Techniques for PRA

The main techniques used for probabilistic risk assessment are fault tree analysis (FTA) and event tree analysis (ETA) [11].

FTA is a graphical relationship among events leading to a "top event" at the apex of the tree. Beginning with the top event, the intermediate events are hierarchically placed at different levels until the required level of detail is reached (the basic events at the bottom of the tree). The interactions between the top event and other events can be generally represented by "OR" or "AND" gates, as shown in **Figure 2(a)** and **(b)**, respectively.

Minimal cut sets (MCSs) of a fault tree are the combinations of basic events which are the shortest pathways that lead to the top event. MCSs are used for qualitative and quantitative assessments of fault trees and can be identified with support of Boolean algebra, specialized algorithms, or computer codes [12]. The probability of the top event can be assessed if the probability values or probability density functions (pdfs) of the basic events are available, using the identified MCSs. For instance, using the set theory concepts [13], the probability equations of the two FTs in **Figure 2(a)** and **(b)** can be expressed by Eqs. (2) and (3), respectively:

$$P(A\ or\ B) = P(A \cup B) = P(A) + P(B) - P(A \cap B), \tag{2}$$

$$P(A\ and\ B) = P(A \cap B) = P(A|B)\ P(B) = P(B|A)\ P(A), \tag{3}$$

where *P(A)* and *P(B)* are the independent probabilities of the basic events and *P(A/B)* and *P(B/A)* are the conditional (dependent) probabilities.

ETA is also a graphical logic model that identifies and quantifies possible outcomes (accident scenarios) following an undesired initiating event [14]. It provides systematic analysis of the time sequence of intermediate events (e.g., success or failure of defense-in-depth layers, as protective system or operator interventions), until an end state is reached. Consequences can be direct (e.g., fires, explosions) or indirect (e.g., domino effects on adjacent plants or environmental consequences).

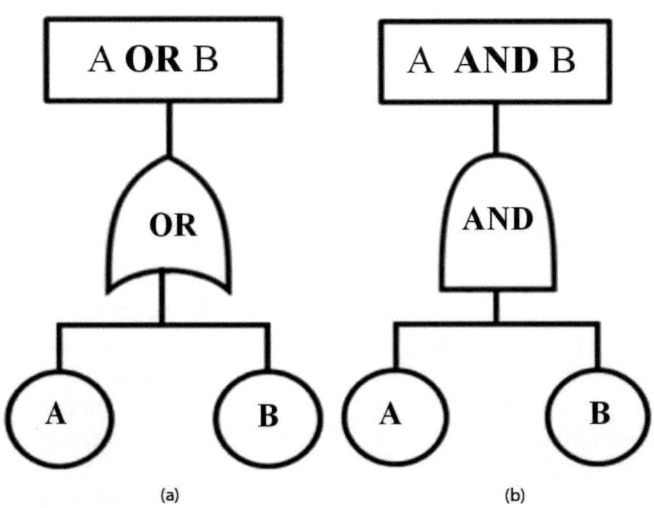

Figure 2.
Intermediate events connected by "OR" (a) and "AND" (b) gates in a fault tree.

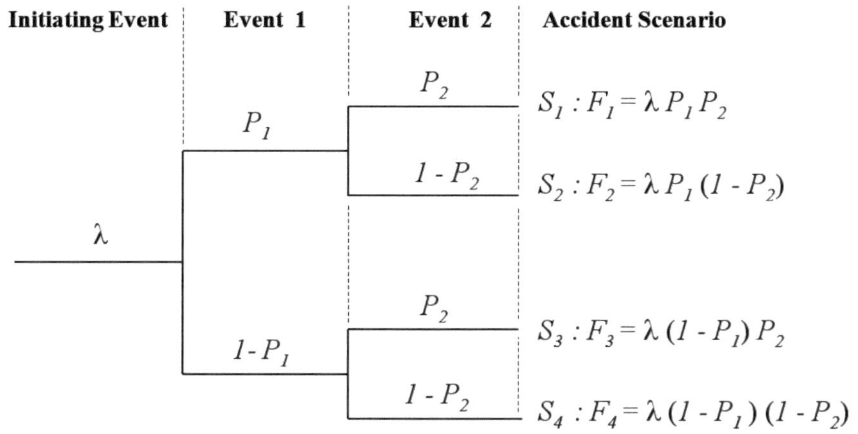

Figure 3.
Sequence of events in an event tree leading to different accident scenarios.

Figure 3 shows an example of an event tree construction, starting with the initiating event of frequency of occurrence, λ, where P_1 and P_2 are the probabilities of subsequent events (event 1 and event 2) leading to the possible scenarios S_1, S_2, S_3, and S_4, with frequencies F_1, F_2, F_3, and F_4, respectively, each one with different consequences. If the success and the failure of each event are mutually exclusive (binary trees) and the probabilities of event occurrence are independent of each other, the frequency of each scenario is calculated as shown in **Figure 3**.

2.3 Uncertainty sources in PRA

Many types of data must be collected and treated for use in PRAs in order to quantify the accident scenarios and accident contributors. Data include, among others, component reliability and failure rates, repair times, initiating event probabilities, human error probabilities, and common cause failure (CCF) probabilities. These data are usually represented by uncertainty bounds or probability density functions, measuring the degree of knowledge or confidence in the available data.

Uncertainties can be highly significant in risk-based decisions and are important for establishing research priorities after a PRA process. For well-understood basic events for which a substantial experience base exists, the uncertainties may be small. When data from experience are limited, the probability of basic events may be highly uncertain, and even knowing that a given probability is small, most of the time one does not know how small it is.

The development of scenarios in a PRA introduces uncertainties about both consequences and probabilities. Random changing of physical processes is an example of stochastic uncertainties, while the uncertainties due to lack of knowledge about these processes are the epistemic uncertainties. Component failure rates and reliability data are typically uncertain, sometimes because unavailability of information and sometimes because doubts about the applicability of available data.

PRA of complex engineering systems such as those in nuclear power plants (NPPs) and chemical plants usually exhibits uncertainties arising from inadequate assumptions, incompleteness of modeling, CCF and human reliability issues, and lack of plant-specific data. For this type of facility, the major of sources of uncertainties are [15]:

- Uncertainties in input parameters—parameters of the models (e.g., FTs and ETs) for estimating event probabilities and assessing magnitude consequences are not exactly known because of the lack of data, variability of plants, processes or components, and inadequate assumptions.

- Modeling uncertainty—inadequacy of conceptual, mathematical, numerical, and computational models.

- Uncertainty about completeness—systematic expert reviewing can minimize the difficulties in assessing or quantifying this type of uncertainty.

The main focus of this work is the treatment of uncertainties regarding numerical values of the parameters used in fault and event trees in the scope of PRA and their propagation in these models. If a probability density function (pdf) is provided for the basic events (e.g., normal, log-normal, or triangular), a pdf or confidence bounds can be obtained for an FT top event or an ET scenario sequence.

3. Methods of uncertainty propagation used in PRA

There are several available methods for propagating uncertainties such as analytical methods (method of moments and Fenton-Wilkinson (FW) method), Monte Carlo simulation, Wilks method (order statistic), and fuzzy set theory. They are different from each other, in terms of characterizing the input parameter uncertainty and how they propagate from parameter level to output level [16].

The analytical methods consist in obtaining the distribution of the output of a model (e.g., fault or event trees) starting from probability distribution of input parameters. An exact analytical distribution of the output however can be derived only for specific models such as normal or log-normal distributions [17].

The Fenton-Wilkinson (FW) method is a kind of analytical technique of approximating a distribution using log-normal distribution with the same moments. It is a moment-matching method for obtaining an exact analytical distribution for the output (closed form). This kind of closed form is helpful, when more detailed uncertainty analyses are required, for instance, in parametric studies involving uncertainty importance assessments, which require re-estimating the overall uncertainty distribution many times [18].

The method of moments is another kind of analytical method where the calculations of the mean, variance, and higher order moments are based on approximate models (generally using Taylor series). As the method is only an approximation, when the variance in the input data are large, higher order terms in the Taylor expansion have to be included. This introduces much more complexity in the analytical model, especially for complex original models, as in the case of PRAs [19].

The Monte Carlo simulation estimates the output parameter (e.g., probability of the top event of an FT) by simulating the real process and its random behavior in a computer model. It estimates the output occurrence by counting the number of times an event occurs in simulated time, starting to sample the pdf from the input data [20].

The fuzzy set theory is used when empirical information for input data are limited and probability theory is insufficient for representing all type of uncertainties. In this case, the so-called possibility distributions are subjectively assigned to input data, and fuzzy arithmetic is carried out. For uncertainty analysis in FTAs, instead of assuming the input parameter as a random variable, it is considered as a fuzzy number, and the uncertainty is propagated to the top event [21].

The Wilks method is an efficient sampling approach, based on order statistics, which can be used to find upper bounds to specified percentiles of the output distribution. Order statistics are statistics based on the order of magnitudes and do not need assumptions about the shape of input or output distributions. According to the authors' knowledge, this method has been of little use in the field of reliability modeling and PRA, although it is used in other aspects of NPP safety, such as uncertainty in input parameters associated with the loss-of-coolant accident (LOCA) phenomena [22].

The mentioned methods for uncertainty propagation have many differences and similarities, advantages and disadvantages, as well as benefits and limitations. **Table 1** summarizes a comparison of these methods.

A brief discussion about the comparison of the mentioned methods is given as follows.

The method of moments is an efficient technique that does not require the specification of the probabilistic distributions of the basic event probabilities. It is difficult to be applied to complex fault trees with many replicated events [23]. This can be solved with the use of computer codes that automatically get the minimal cut sets (MCSs) of the fault trees. It is a simple method, easily explainable and suited for screening studies, due to inherent conservatism and simplicity [24].

The Monte Carlo simulation is computationally intensive for large and complex systems and requires pdf of input data. It has the disadvantage of not readily revealing the dominant contributors to the uncertainties. With current computer technology and availability of user-friendly software for Monte Carlo simulation, computational cost is no longer a limitation.

The fuzzy set theory does not need detailed empirical information like the shape of distribution, dependencies, and correlations. Fuzzy numbers are a good representation of uncertainty when empirical information is very scarce. It is inherently conservative because the inputs are treated as fully correlated [25].

The Fenton-Wilkinson (FW) method improves the understanding of the contributions to the uncertainty distribution and reduces the computational costs involved, for instance, in conventional Monte Carlo simulation for uncertainty

Method	Propagation technique	Benefits	Limitations
Method of moments	Analytical (probability theory and statistics)	Conceptually simple and does not require the specification of pdf of input data	Difficult to apply for complex systems and large fault trees
Monte Carlo simulation	Simulation	Estimates are closes to exact solutions, especially for simple and small systems	Computationally intensive for large and complex systems. Requires pdf of input data and does not reveal contributors to the uncertainty
Fuzzy set theory	Fuzzy arithmetic	It does not require detailed information of pdf. Suited when empirical information is very scarce	It is inherently conservative because the inputs are treated in a fully correlated way
Fenton-Wilkinson (FW) method	Analytical (closed-form approximation)	Improves understanding of contributions to uncertainties and has low computational costs	Closed form for top events is not easily obtained. Applicable only to log-normal distribution. Estimates are most accurate in the central range
Wilks method	Order statistics	Conservative and computationally inexpensive	Low accuracy in low tails of the distributions

Table 1.
Comparison of methods for uncertainty propagation.

estimation. It is applicable only when the uncertainties in the basic events of the model are log-normally distributed. FW estimates are most accurate in the central range, and the tails of the distributions are poorly represented. The Wilks method requires relatively few samples and is computationally inexpensive. It is useful for providing an upper bound (conservative) for the percentiles of the uncertainty distribution. However, its calculated values are less accurate than the FW estimates over practically the entire range of the distribution. For both Wilks and FW methods, the greatest errors are found in the low tails of the distributions, but in almost all reliability applications the high tails are of more interest than the low tails [26].

4. Method of moments for uncertainty propagation in FTA and ETA

The method of moments uses first and second moments of the input parameters (mean and variance) to estimate the mean and variance of the output function using propagation of variance or coefficient of variation. As a measure of uncertainty, the coefficient of variation is defined as a ratio of the standard deviation to the mean, which indicates the relative dispersion of uncertain data around the mean. The uncertainty measure is a readily interpretable and dimensionless measure of error, differently for standard deviation, which is not dimensionless [27].

In PRA, the method of moments can be used to propagate the uncertainties of the inputs (i.e., event probabilities) and propagate the uncertainty for the outputs. The probability density functions (pdfs) for the inputs can be estimated from reliability data of gathered components or from historical records of undesired events. Hypothesizing that the events (or basic events) are independent, probabilistic approaches for propagating uncertainties in FTs and ETs are given as follows in Sections 4.1 and 4.2, respectively [28].

4.1 Method of moments applied to FTA

The uncertainty propagation in a fault tree begins with the propagation of uncertainties of basic events through "OR" and "AND" gates, until it reaches the top event. The fault tree should be represented by MCSs in order to avoid direct dependence between intermediate events, facilitating probabilistic calculations.

For an "OR" gate of a fault tree, the probability of the output event, P_{or}, is given by Eq. (4):

$$P_{or} = 1 - \prod_{i=1}^{n} (1 - P_i), \tag{4}$$

where P_i denotes the probability of ith (i = 1, 2, 3, ..., n) independent events (or basic events) and n is the number of input events.

The uncertainty propagation through the "OR" gate is given by Eq. (5) that calculates the coefficient of variation of output, C'_{or}, as function of the coefficients of variation of inputs, C'_i, according to Eqs. (6) and (7) [29]:

$$1 + C'^2_{or} = \prod_{i=1}^{n} (1 + C'^2_i), \tag{5}$$

$$C'_{or} = \frac{s_{or}}{1 - P_{or}}, \tag{6}$$

$$C_i' = \frac{s_i}{1 - P_i}, \tag{7}$$

where s_i denotes the standard deviations of ith (i = 1, 2, 3, …, n) input, n is the number of input events, and s_{or} is the standard deviation of the output of "OR" gate.

For an "AND" gate of a fault tree, the probability of output event, P_{and}, is given by Eq. (8):

$$P_{and} = \prod_{i=1}^{n} P_i, \tag{8}$$

where P_i denotes the probability of ith (i = 1, 2, 3, …, n) independent events (or basic events) and n is the number of input events.

The uncertainty propagation through the "AND" gate is given by Eq. (9). It calculates the coefficient of variation of output, C_{and}, as function of the coefficients of variation of inputs, C_i, according to Eqs. (10) and (11) [29]:

$$1 + C_{and}^2 = \prod_{i=1}^{n} \left(1 + C_i^2\right), \tag{9}$$

$$C_{and} = \frac{s_{and}}{P_{and}}, \tag{10}$$

$$C_i = \frac{s_i}{P_i}, \tag{11}$$

where s_i denotes the standard deviations of ith (i = 1, 2, 3, …, n) input, n is the number of input events, and s_{and} is the standard deviation of output of the "AND" gate.

4.2 Method of moments applied to ETA

Uncertainty propagation in an event tree is similar (or analogous) to uncertainty propagation of an "AND" gate of a fault tree. The frequency of occurrence of each accident scenario, F_{seq}, is given by Eq. (12),

$$F_{seq} = \lambda \times \prod_{i=1}^{n} P_i, \tag{12}$$

where λ is the frequency of occurrence of the initiating event and P_i denotes the probabilities of ith (i = 1, 2, 3, …, n) subsequent independent events leading to the accident scenario and n is the number of input events. These values can be obtained from fault trees constructed for each ith event or system failure of the event tree.

The uncertainty propagation through the accident sequence is given by Eq. (13) that provides the coefficient of variation of accident sequence, C_{seq}, as function of the coefficients of variation of subsequent events, C_i, according to Eqs. (14) and (15), respectively:

$$1 + C_{seq}^2 = \prod_{i=1}^{n} \left(1 + C_i^2\right), \tag{13}$$

$$C_{seq} = \frac{s_{seq}}{F_{seq}}, \tag{14}$$

$$C_i = \frac{s_i}{P_i}, \tag{15}$$

where s_i denotes the standard deviations of ith (i = 1, 2, 3, ..., n) subsequent event of the sequence, n is the number of input events, and s_{seq} is the standard deviation of the accident sequence.

4.3 Propagation of log-normal distributions

Many uncertainty distributions associated with the basic events of fault trees (reliability or failure probability data) often can be approximated in reliability and safety studies by log-normal functions. If a random variable $\ln(x)$ has a normal distribution, the variable x has then a log-normal distribution. The log-normal probability density function (pdf), $f(x)$ is then given by Eq. (16) [30]:

$$f(x) = \frac{1}{x\sigma\sqrt{2\pi}} \exp\left(\frac{-(\ln(x) - \mu)^2}{2\sigma^2}\right), \tag{16}$$

where μ and σ are the mean and the standard deviation of $\ln(x)$, respectively (i.e., these are the parameters of the "underlying" normal distribution).

The error factor, *EF*, of a log-normal pdf is defined as Eq. (17):

$$EF = \frac{\chi_{95}}{\chi_{50}} = \frac{\chi_{50}}{\chi_5}, \tag{17}$$

where χ_{95}, χ_{50}, and χ_5 are the 95th, 50th (median), and 5th percentiles, respectively.

EF is often used as an alternative to the standard deviation of "underlying" normal distribution, σ, for characterizing the spread of a log-normal distribution, and these two quantities are related by Eq. (18):

$$EF = \exp(1.645\,\sigma). \tag{18}$$

The mean, P, and standard deviation, s, of the log-normal variable, x, can be given by the following Eqs. (19) and (20), respectively:

$$P = \exp\left(\mu + \frac{\sigma^2}{2}\right), \tag{19}$$

$$s = \sqrt{\exp(2\mu + \sigma^2)[\exp(\sigma^2) - 1]} \tag{20}$$

Eqs. (4)–(20) are used for uncertainty propagation of log-normal pdf in fault and event trees, as illustrated in the following examples.

5. Illustrative examples

In order to validate the proposed approach for implementing the method of moments, two cases were tested.

5.1 Case study 1

The first case, taken from Chang et al. [8], introduces a fault tree (**Figure 4**) describing a generic top event "system failure," T, with seven basic events (X(1) to

X(7)), characterized by the log-normal distributions. This simple example was chosen in order to compare the results of the method of moments with the uncertainty propagation analyses using Monte Carlo simulation.

The log-normal distributions assigned to the basic events (represented by median and mean values of probabilities, error factors, and standard deviations) are shown in **Table 2**. An analysis of the fault tree shows that its minimal cut sets (MCSs) are X(1), X(6), X(7), X(2)X(4), X(2)X(5), X(3)X(4), and X(3)X(5), which are used to estimate the top event probability and propagate the uncertainties. The application of the method of moments is carried out in a bottom-up approach. Starting from basic events of the fault tree, the coefficients of variation of the intermediate events are estimated using Eqs. (4)–(7) for "OR" gates and Eqs. (8)–(11) for "AND" gates. This procedure is repeated interactively until the top event is reached, and its standard deviation is obtained. Considering that, in the same way as the basic events, the top event has also a log-normal distributions, Eqs. (16)–(20) are used to estimate the 5th percentile, median, and 95th percentile for the top event, as shown in **Table 3**. These estimates are slightly lower than the values obtained by Chang et al. [8] with the Monte Carlo simulation (percent

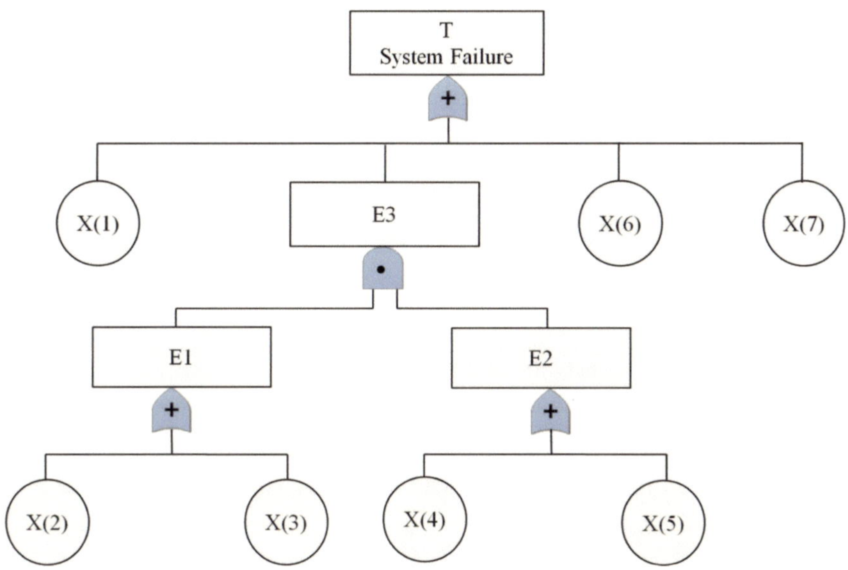

Figure 4.
Fault tree analysis for a generic top event "system failure" (adapted from Chang et al. [8]).

Basic events	Median of log-normal pdf (χ_{50})	*EF* of log-normal pdf	Mean of log-normal pdf *(P)*	Standard deviation of log-normal pdf (*s*)
X(1)	1.00×10^{-3}	3	1.25×10^{-3}	9.37×10^{-4}
X(2)	3.00×10^{-2}	3	3.75×10^{-2}	2.81×10^{-2}
X(3)	1.00×10^{-2}	3	1.25×10^{-2}	9.37×10^{-3}
X(4)	3.00×10^{-3}	3	3.75×10^{-3}	2.81×10^{-3}
X(5)	1.00×10^{-2}	3	1.25×10^{-2}	9.37×10^{-3}
X(6)	3.00×10^{-3}	3	3.75×10^{-3}	2.81×10^{-3}
X(7)	1.00×10^{-3}	3	1.25×10^{-3}	9.37×10^{-4}

Table 2.
Basic event distribution for a generic top event "system failure" (χ_{50} and EF values were taken from Ref. [8]).

Method	5th percentile	Median	95th percentile
Monte Carlo simulation[1]	4.15×10^{-3}	8.02×10^{-3}	1.64×10^{-2}
Method of moments[2]	3.99×10^{-3}	7.95×10^{-3}	1.58×10^{-2}
% difference	−3.8%	−0.9%	−3.5%

[1]*Ref. [8].*
[2]*Current work.*

Table 3.
Comparison of top event probabilities obtained by Monte Carlo simulation and by method of moments.

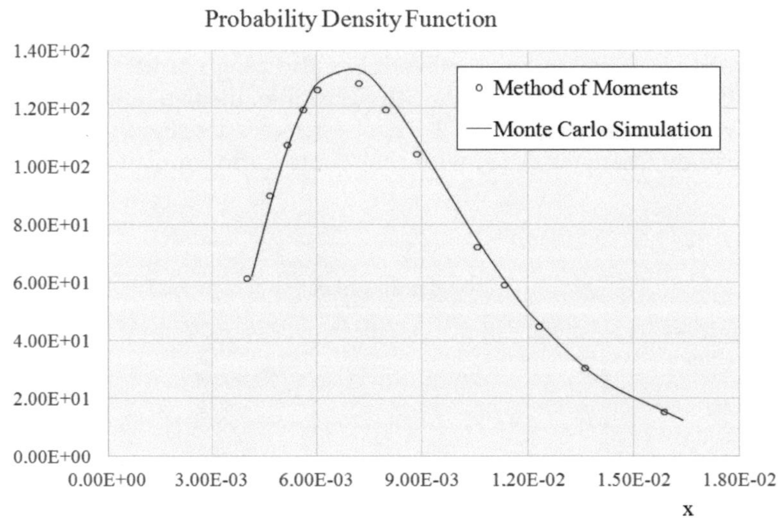

Figure 5.
Comparison of pdf obtained by method of moments and by the Monte Carlo simulation for the top event of ***Figure 4****.*

difference less than 4%). This good agreement can also be verified through the probability density function (obtained with Eq. (16)), as shown in **Figure 5**.

5.2 Case study 2

The second case study illustrates the application of the method of moments for assessing the uncertainty of a fault tree taken from a probabilistic safety analysis of a nuclear power plant (NPP). The fault tree shown in **Figure 6** was constructed using MCSs and basic event distributions provided by El-Shanawany et al. [26]. It represents a fault tree analysis for the top event "nuclear power plant core melt," taking into account loss of off-site and on-site power systems and failure of core residual heat removal. The basic events A, B, C, D, E, F, G, H, I, J, K, L, and M are related to off-site power system failure, operator errors, emergency diesel generators (EDGs) failures, pump failures, and common cause failures (CCFs). A detailed description of each one of these basic events is given in the caption of **Figure 6**. An accurate logical analysis of this drawn fault tree can demonstrate that its MCSs are ABC, ABD, ABE, ABF, ABH, ABI, ABJ, AFG, and AKLMH, which describes the illustrative example analyzed in the literature.

The log-normal distributions assigned to the basic events (represented by mean values of probabilities, error factors, and standard deviations) are shown in **Table 4**. Such distributions are also used in Ref. [26], to compare the results of this current work, using the method of moments, with the analyses of uncertainty propagation using Wilks method, Monte Carlo simulation, and Fenton-Wilkinson (FW) method.

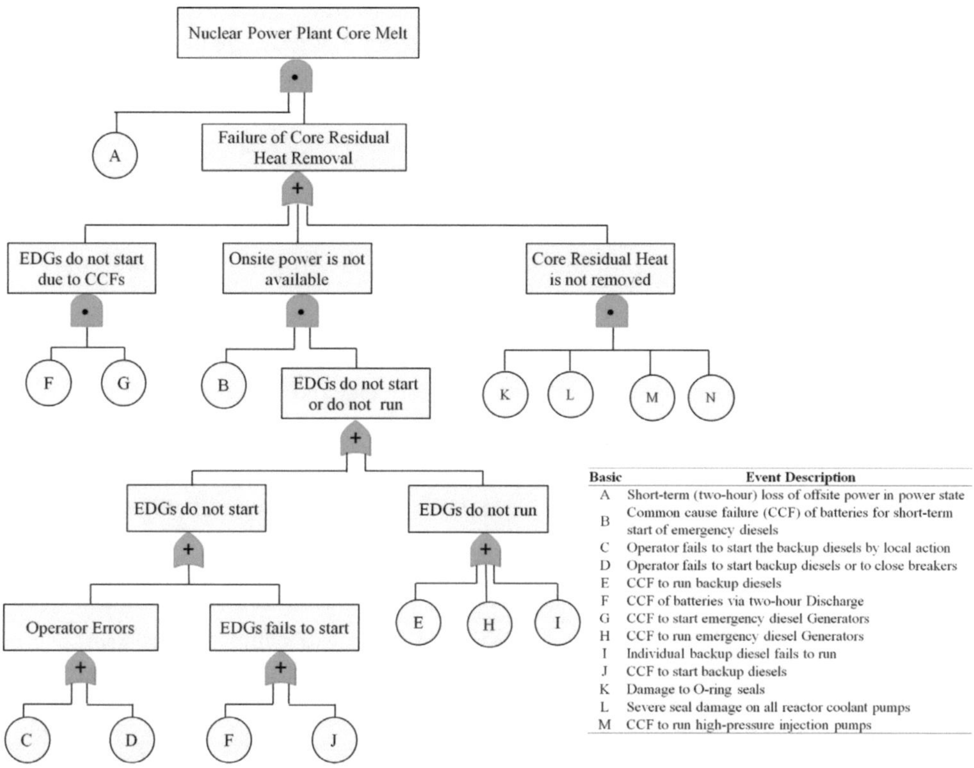

Figure 6.
Fault tree analysis for a nuclear power plant core melt.

Basic events	Mean of log-normal pdf *(P)*	Error factor of log-normal pdf *(EF)*	Standard deviation of log-normal pdf *(s)*
A	6.00×10^{-2}	5	7.60×10^{-2}
B	6.60×10^{-6}	5	8.36×10^{-6}
C	1.00×10^{-2}	5	1.27×10^{-2}
D	2.13×10^{-3}	5	2.70×10^{-3}
E	8.33×10^{-4}	5	1.06×10^{-3}
F	5.20×10^{-5}	5	6.59×10^{-5}
G	6.10×10^{-5}	5	7.73×10^{-5}
H	4.20×10^{-5}	5	5.32×10^{-5}
I	1.58×10^{-3}	5	2.00×10^{-3}
J	1.00×10^{-4}	5	1.27×10^{-4}
K	9.00×10^{-2}	5	1.14×10^{-1}
L	1.00×10^{-1}	5	1.27×10^{-1}
M	1.20×10^{-4}	5	1.52×10^{-4}

Table 4.
Basic event distribution for illustrative example (P and EF values were taken from Ref. [26]).

The application of the method of moments is carried out in a similar way as in the first case study. Considering that the top event is also log-normally distributed, its 5th percentile, median, and 95th percentile are estimated. As can be seen in **Table 5**, the median values of the method of moments show a good agreement with Wilks method and are 25.8% and 20.4% greater than the results of Monte Carlo simulation and FW method, respectively. This is also illustrated in **Figure 7**, where the cumulative distribution function obtained by method of moments is compared with the data in the mentioned literature [26]. As can be seen, the results of the method of moments agree reasonably with the Wilks method, being slightly lower, moving toward the analyses of uncertainty propagation using Monte Carlo simulation, which is considered for many purposes to be close to the exact solution for simple models.

Overall, uncertainty propagation using the method of moments in fault trees, as shown in the two case studies, or in event trees, is quite simple in small systems and does not require the specification of probability density functions of basic events but only their means and standard deviations. For more complex systems and large fault and event trees, computer implementation of the described bottom-up approach can be performed, for instance, using specialized computer software for obtaining the minimal cut sets and quantitatively assessing the top event

Method	5th percentile	Median	95th percentile	% difference of median of method of moments
Monte Carlo simulation[1]	8.80×10^{-11}	1.55×10^{-9}	2.10×10^{-8}	25.8%
Fenton-Wilkinson[1]	1.26×10^{-10}	1.62×10^{-9}	2.08×10^{-8}	20.4%
Wilks[1]	1.85×10^{-10}	1.95×10^{-9}	2.46×10^{-8}	0.0%
Moments[2]	1.65×10^{-10}	1.95×10^{-9}	2.31×10^{-8}	—

[1]*Ref. [26].*
[2]*Current work.*

Table 5.
Comparison of core melt frequency obtained by the method of moments with data from literature.

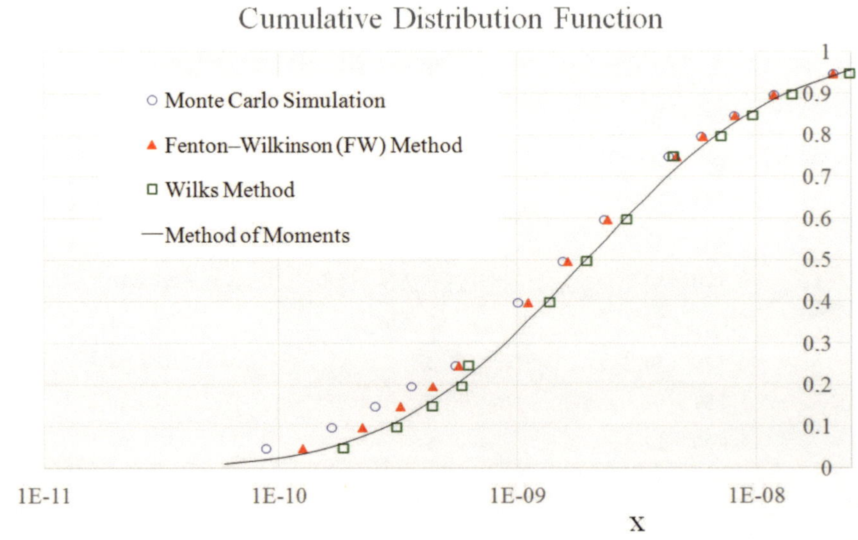

Figure 7.
Comparison of cumulative distribution function for core melt frequency obtained by the method of moments with data from literature [26].

probabilities [31], as well as matrix computations for obtaining the standard deviations along the trees, as proposed by Simões Filho [32].

6. Final remarks

This work addresses the uncertainty propagation in fault and event trees in the scope of probabilistic risk assessment (PRA) of industrial facilities. Given the uncertainties of the primary input data (component reliability, system failure probabilities, or human error rates), the method of moments is proposed for the evaluation of the confidence bounds of top event probabilities of fault trees or event sequence frequencies of event trees. These types of analyses are helpful in performing a systematic PRA uncertainty treatment of risks and system reliabilities associated with complex industrial facilities, mainly in risk-based decision-making.

Two illustrative examples using the method of moments for carrying out the uncertainty propagation in fault trees are presented, and their results are compared with available analyses in literature using different uncertainty assessment approaches. The method of moments proved to be conceptually simple to be used. It confirmed findings postulated in literature, when dealing with simple and small systems. More complex systems will require the support of specialized reliability and risk assessment software, in order to implement the proposed approach.

Acknowledgements

The authors would like to thank the following institutions, which sponsored this work: Development Center of Nuclear Technology/Brazilian Nuclear Energy Commission (CDTN/CNEN) and Brazilian Innovation Agency (FINEP).

Conflict of interest

The authors are the only responsible for the printed material included in this paper.

Author details

Vanderley de Vasconcelos*, Wellington Antonio Soares,
Antônio Carlos Lopes da Costa and Amanda Laureano Raso
Development Center of Nuclear Technology/Brazilian Nuclear Energy Commission, CDTN/CNEN, Belo Horizonte, Brazil

*Address all correspondence to: vasconv@cdtn.br

References

[1] Reinert JM, Apostolakis GE. Including model uncertainty in risk-informed decision making. Annals of Nuclear Energy. 2006;**33**:354-369

[2] U.S. Nuclear Regulatory Commission (USNRC). WASH-1400: Reactor Safety Study (NUREG-75/014). Washington, DC: USNRC; 1975

[3] Innal F, Mourad C, Bourareche M, Antar AS. Treatment of uncertainty in probabilistic risk assessment using Monte Carlo analysis. In: Proceedings of the 3rd International Conference on Systems and Control; 29–31 October 2013. Algiers, Algeria

[4] Christensen FM, Andersen O, Duijm NJ, Harremoës P. Risk terminology: A platform for common understanding and better communication. Journal of Hazardous Materials. 2003;**A103**: 181-203

[5] Stamatelatos M. Probabilistic Risk Assessment Procedures Guide for NASA Managers and Practitioners: Version 1.1. Washington, DC: Office of Safety and Mission Assurance, NASA Headquarters; 2002

[6] International Standard Organization (ISO). Risk Management: Risk Assessment Techniques (ISO IEC/FDIS 31010). Geneva, Switzerland: ISO/IEC; 2009

[7] U.S. Nuclear Regulatory Commission (USNRC). Fire Dynamics Tools (FDTs): Quantitative Fire Hazard Analysis Methods for the U.S. Nuclear Regulatory Commission Fire Protection Inspection Program (NUREG 1805). Washington, DC: USNRC; 2004

[8] Chang SH, Park JY, Kim MK. The Monte-Carlo method without sorting for uncertainty propagation analysis in PRA. Reliability Engineering. 1985;**10**: 233-243

[9] El-Shanawany AB. Quantification of uncertainty in probabilistic safety analysis [Thesis]. London: Imperial College London, Department of Mechanical Engineering; 2017

[10] Goerlandt F, Reniers G. On the assessment of uncertainty in risk diagrams. Safety Science. 2016;**84**:67-77

[11] Vasconcelos V, Soares WA, Marques RO. Integrated engineering approach to safety, reliability, risk management and human factors. In: Felice F, Petrillo A, editors. Human Factors and Reliability Engineering for Safety and Security in Critical Infrastructures: Decision Making, Theory, and Practice. Cham, Switzerland: Springer International Publishing AG; 2018. pp. 77-107

[12] Reliasoft. System Analysis Reference: Reliability, Availability and Optimization. Tucson AZ: ReliaSoft Corporation; 2015

[13] Vesely WE, Goldberg FF, Roberts NH, Haasl DF. Fault Tree Handbook (NUREG-0492). Washington, DC: USNRC, Office of Nuclear Regulatory Research; 1981

[14] U.S. Nuclear Regulatory Commission (USNRC), PRA Procedures Guide. A Guide to the Performance of Probabilistic Risk Assessments for Nuclear Power Plants (NUREG/CR-2300). Washington, DC: USNRC; 1983

[15] Durga Rao K, Kushwaha HS, Verma AK, Srividya A. Epistemic uncertainty propagation in reliability assessment of complex systems. International Journal of Performability Engineering. 2007; **3**(4):71-84

[16] Suresh PV, Babar AK, Venkat Raj V. Uncertainty in fault tree analysis: A fuzzy approach. Fuzzy Sets and Systems. 1996;**83**:135-141

[17] Ulmeanu AP. Analytical method to determine uncertainty propagation in fault trees by means of binary decision diagrams. IEEE Transactions on Reliability. 2012;**61**(1):84-94

[18] Dezfuli H, Modarres M. Uncertainty analysis of reactor safety systems with statistically correlated failure data. Reliability Engineering. 1985;**11**:47-64

[19] Cheng D. Uncertainty analysis of large risk assessment models with applications to the railway safety and standards board safety risk model [Thesis]. Glasgow: University of Strathclyde, Department of Management Science; 2009

[20] Raychaudhuri S. Introduction to Monte Carlo simulation. In: Proceedings of the 2008 Winter Simulation Conference; 7–10 December 2008. Miami, USA

[21] Ferdous R, Khan F, Sadiq R, Amyotte P, Veitch B. Fault and event tree analyses for process systems risk analysis: Uncertainty handling formulations. Risk Analysis. 2011;**31**(1): 86-107

[22] Lee SW, Chung BD, Bang YS, Bae SW. Analysis of uncertainty quantification method by comparing Monte Carlo method and Wilks' formula. Nuclear Engineering and Technology. 2014;**46**(4):481-488

[23] Ahmed S, Metcalf DR, Pegram JW. Uncertainty propagation in probabilistic risk assessment: A comparative study. Nuclear Engineering and Design. 1981; **68**:1-31

[24] Rushdi AM, Kafrawy KF. Uncertainty propagation in fault tree analyses using an exact method of moments. Microelectronics and Reliability. 1988;**28**(6):945-965

[25] Ferdous R. Quantitative risk analysis in an uncertain and dynamic environment [Thesis]. Newfoundland, Canada: Faculty of Engineering & Applied Science, Memorial University of Newfoundland; 2011

[26] El-Shanawany AB, Ardron KH, Walker SP. Lognormal approximations of fault tree uncertainty distributions. Risk Analysis. 2018;**38**(8):1576-1584

[27] Apostolakis G, Lee YT. Methods for the estimation of confidence bounds for the top-event unavailability of fault trees. Nuclear Engineering and Design. 1977;**41**:411-419

[28] Bier VM. Uncertainty analysis as applied to probabilistic risk assessment. In: Covello VT, Lave LB, Moghissi AA, Uppuluri VRR, editors. Uncertainty in Risk Assessment, Risk Management, and Decision Making. New York: Plenum Press; 1987. pp. 469-478

[29] Ahn K. On the use of coefficient of variation for uncertainty analysis in fault tree analysis. Reliability Engineering and System Safety. 1995;**47**: 229-230

[30] Reliasoft. Life Data Analysis Reference. Tucson, AZ: ReliaSoft Corporation; 2015

[31] Misra KB. Handbook of Performability Engineering. Cham, Switzerland: Springer International Publishing AG; 2008

[32] Simões Filho S. Análise de árvore de falhas considerando incertezas na definição dos eventos básicos [thesis]. Rio de Janeiro, Brazil: Faculdade de Engenharia Civil, Universidade Federal do Rio de Janeiro; 2006

Chapter 8

Reliability Evaluation of Power Systems

Abdullah M. Al-Shaalan

Abstract

Reliability evaluation of electric power systems is an essential and vital issue in the planning, designing, and operation of power systems. An electric power system consists of a set of components interconnected with each other in some purposeful and meaningful manner. The object of a reliability evaluation is to derive suitable measures, criteria, and indices of reliable and dependable performance based on component outage data and configuration. For evaluating generated reliability, the components of interest are the generating units and system configuration, which refer to the specific unit(s) operated to serve the present or future load. The indices used to measure the generated reliability are probabilistic estimates of the ability of a particular generation configuration to supply the load demand. These indices are better understood as an assessment of system-wide generation adequacy and not as absolute measures of system reliability. The indices are sensitive to basic factors like unit size and unit availability and are most useful when comparing the relative reliability of different generation configurations. The system is deemed to operate successfully if there is enough generation capacity (adequate reserve) to satisfy the peak load (maximum demand). Firstly, generation model and load model are convolved (mutually combined) to yield the risk of supply shortages in the system. Secondly, probabilistic estimates of shortage risk are used as indices of bulk power system reliability evaluation for the considered configuration.

Keywords: reliability, outage, availability, energy, power system, systems interconnection

1. Introduction

Reliability is one of the most important criteria, which must be taken into consideration during all phases of power system planning, design, and operation. A reliability criterion is required to establish target reliability levels and to consistently analyze and compare the future reliability levels with feasible alternative expansion plans. This need has resulted in the development of comprehensive reliability evaluation and modeling techniques [1–6]. As a measure of power system reliability evaluation in generation expansion planning and energy production, three fundamental indices are widely adopted and used.

The first reliability index is the loss of load expectation (*LOLE*) which denotes the expected average number of days per year during which the system is being on outages, i.e., load exceeds the available generating capacity.

The second index is the expected demand not supplied (ϵDNS) which measures the size of load that has been lost due to the severe outages occurrence.

The third index is the expected energy not supplied (ϵENS), which is defined as the expected size of energy not being supplied by the generating unit(s) residing in the system during the period considered due to capacity deficit or unexpected severe power outages [7, 8].

The implementations of these indices are now increasing since they are significant in physical and economic terms. Compared with generation reliability evaluation, there are also reliability indices related and pertinent to network (transmission and distribution) reliability evaluation.

There are two basic concepts usually considered in network reliability, namely, violation of quality and violation of continuity.

The first criterion considers violation of voltage limits and violation of line rating or carrying capacity, and the second criterion assumes that lines are of infinite capacity.

The transmission and distribution networks can be analyzed in a similar manner to that used in generation reliability evaluation, that is, the probability of not satisfying power continuity. This would give frequency and duration in network evaluation a simplification that is necessary. Provided the appropriate component reliability indices are known, it is relatively simple to calculate the expected failure rate (λ) of the system, the average duration of the outage (r), and the unavailability (U). To do this, the values of λ, r and U are required for each component of the system [9–11].

2. Types of system outages and deficits

A bulk generation model must consider the size of generation reserve and the severe outage(s) occurrences. An outage in a generating unit results in the unit being removed from service in order to be repaired or replaced. Such outages can compromise the ability of the system to supply the load and, hence, affect system reliability. An outage may or may not cause an interruption of service depending on the margins of generation provided. Outages also occur when the unit undergoes maintenance or other planned works necessary to keep it operating in good condition. The outages can be classified into two categories:

- A planned outage that results when a component is deliberately taken out of service, usually for purposes of preventive repair or planned maintenance
- A forced outage that results from sudden and emergency conditions, forcing the generating unit to be taken out of service

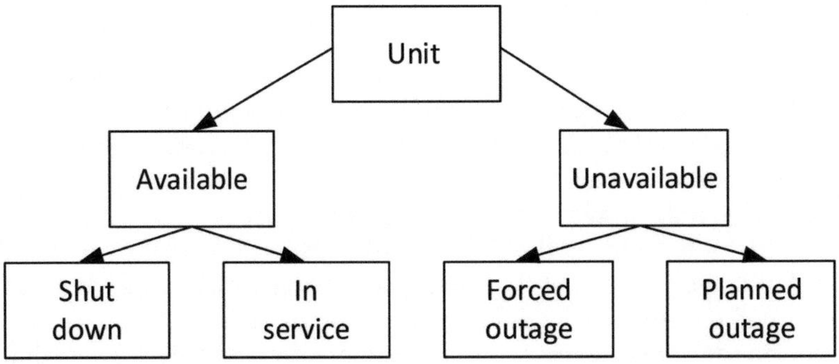

Figure 1.
Generating unit probable states.

The status of a generating unit is described as morphing into one of the several possible states, as shown in **Figure 1**.

To investigate the effect of a unit on system generation reliability, it is imperative to know its probability of residing in each state as in **Figure 1**. Hence, the following section introduces some basic probability concepts.

3. Introduction to power system reliability evaluation

3.1 Availability (AV) and forced outage rate (FOR)

Experience has shown that no machine is so reliable and dependable that it is available in successful operating condition all the time. That means that the machine needs to be off service (out of service) for maintenance or it may be off due to some other problems affecting its operation (see **Figure 1**). As such, an off-service status includes planned outages and forced outages. Planned outages (scheduled outages) are the ones when (a) unit(s) is purposely shutdown or taken out of service for maintenance or replacement. Forced outages are defined as the ones when (a) unit(s) is out of service due to failure (also called unscheduled or unplanned outage). The last one is the most severe and important factor in power system planning and operation and can be defined as

$$\text{Forced outage rate (FOR)} = \frac{\text{sum of time unit is being out of service}}{\text{Total time considered for unit service}} \tag{1}$$

$$\text{FOR} = \frac{t_1 + t_2 + t_3}{\text{Total time}} \tag{2}$$

Also, availability can be defined as

$$A\text{vailability (AV)} = \frac{\text{Time unit is being in service}}{\text{Total time considered for unit service}} \tag{3}$$

and $$\text{AV} + \text{FOR} = 1.$$

This can be seen in **Figure 2** as follows

The two terms "availability and forced outage rate" represent the probability of successful and failure event occurrence. According to the probability theory, it is known that the product $AV_1 \times AV_2$ represents the probability that both unit 1 and unit 2 are simultaneously in operation during a specified interval of time, and, also, $AV_1 \times AV_2 \times AV_3$ means 1 and 2 and 3 are in operation at the same time, and $FOR_1 \times FOR_2 \times FOR_3$ means that units 1, 2, and 3 are out of service in the same time.

Also, $AV_1 \times FOR_2$ means the probability that unit 1 is available (in service) and unit 2 is unavailable (out of service) in the same time.

For system generation reliability evaluation (including system expansion planning and/or systems interconnection), two models, namely, capacity model and load model, are needed; these are demonstrated and elaborated in the next two sections.

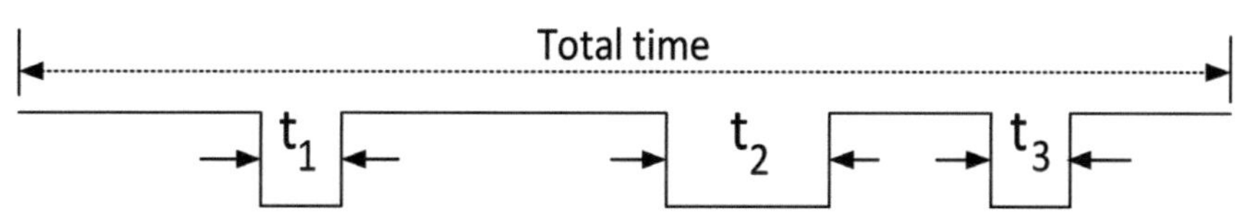

Figure 2.
Unit being available and unavailable.

3.2 Capacity model

The capacity model is known as the "Capacity Outage Probability Table (COPT)" that contains all capacity states (available and non-avoidable) in an ascending order of outage magnitude. Each outage (capacity state) is multiplied by its associated probability. If the system contains identical units, the binomial distribution can be used [12].

3.3 Load model

The load model is known as the "load duration curve (LDC)" which is the most favorable one to be used instead of the regular load variation curve. There are some facts about the LDC that should be realized and can be summarized as follows:

a. The LDC is an arrangement of all load levels in a descending order of magnitude.

b. The area under the LDC represents the energy demanded by the system (consumed).

c. LDC can be used in economic dispatching, reliability evaluation, and power system planning and operation.

d. It is more convenient to deal with than the regular timely load variation curve.

The above discussion for the load duration curve is depicted in **Figure 3** with all pertinent captions related to it.

3.4 Loss of load expectation (*LOLE*)

The *LOLE* risk index is the most widely accepted and utilized probabilistic method in power generation reliability evaluation for purposes of system expansion and interconnection. The two models, namely, the COPT and the LDC, mentioned

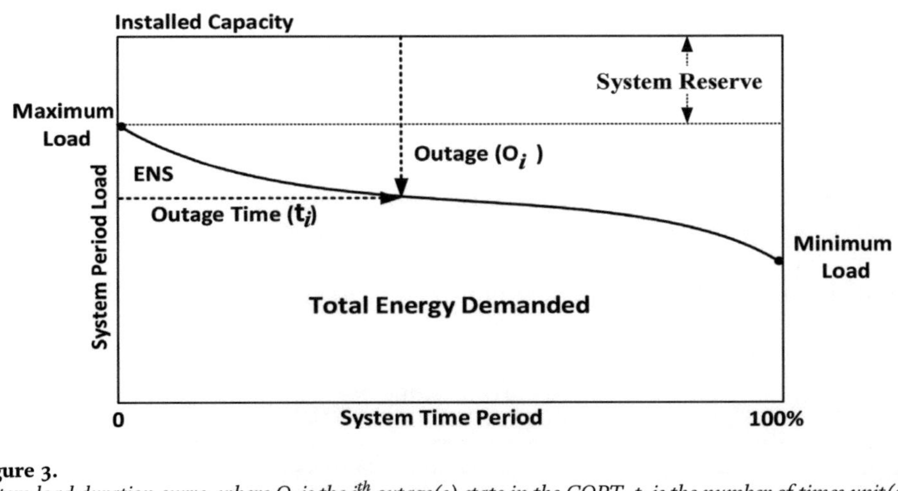

Figure 3.
System load duration curve, where O_i is the ith outage(s) state in the COPT, t_i is the number of times unit(s) is unavailable, P_i is the probability of this i^{th} unavailable, and ENS is the energy not supplied due to severe outage (s) occurrence.

in the preceding sections are convolved (combined) in the process. The unit of the *LOLE* is in days per year (d/y). The *LOLE* evaluation method is expressed in the following mathematical formula:

$$LOLE = \sum_{i=1}^{n} t_i \cdot p_i(o_i)\ days/year\ [L_{max} > Reserve] \tag{4}$$

By observing the above equation, the *LOLE* would be applicable if, and only if, the maximum load (L_{max}) exceeds the system reserve. Consider now:

3.5 Expected demand not supplied (ϵDNS)

In power system planning another reliability index beside the *LOLE* may be required, so as to determine the size and magnitude of the load that has been lost due to severe outages (i.e., when $L_{max} > system\ reserve$), Hence, the ϵDNS can be obtained as follows:

$$\epsilon DNS = \sum_{i=1}^{n} (DNS_i) \cdot p_i\ MW/year \qquad [L_{max} > Reserve] \tag{5}$$

3.6 Expected energy not supplied (ϵENS)

Since power systems are in fact energy systems, the expected energy not supplied index may be deduced as per **Figure 4**. The ϵENS index is used in order to calculate energy sale, which is the real revenue for any electric company.

$$\epsilon ENS = \sum_{i=1}^{n} (ENS_i) \cdot p_i\ MWh/year\ [L_{max} > Reserve] \tag{6}$$

3.7 Energy index of reliability (EIR)

The ratio of expected energy not supplied ($\epsilon \boldsymbol{ENS}$) to the system's total energy demanded ($\boldsymbol{TED}$) can be found as

$$\epsilon ENS_{pu} = \frac{\epsilon ENS}{TED} \tag{7}$$

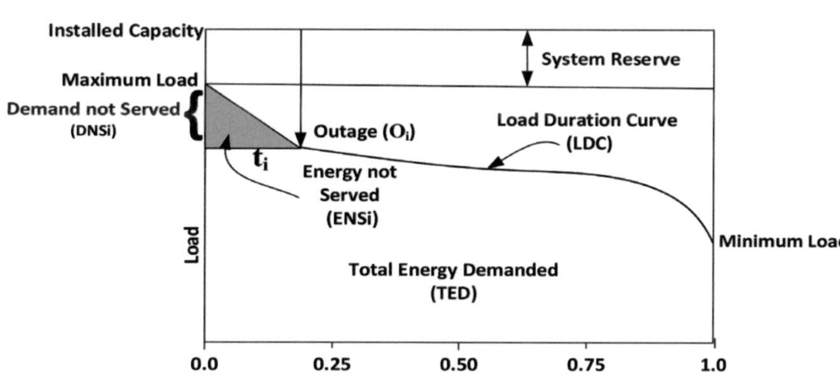

Figure 4.
Load duration curve with energy not served.

This ratio, in fact, is so small because of the small nature of the ϵENS and the large nature of the ***TED***, so, one can deduce another important reliability index called the ***EIR***, which can be expressed as follows

$$EIR = 1 - \epsilon ENS_{pu} \tag{8}$$

4. Energy production evaluation methodology

4.1 Basic concept

The expected energy supplied (ϵES) by the generating units (existing in the system) can be evaluated by using the concept of the expected energy not supplied (ϵENS) described previously. In this method, several factors are taken into consideration:

- Unit forced outage rate (FOR).
- Load duration curve (LDC).
- Capacity-Availability Table (CAT): a table that contains all the capacity states of the units in the system arranged according to their ascending order of availabilities.
- Loading priority levels: implies loading units in accordance to their least operating cost, i.e., operating, first, the most efficient and economical operating units (called the base units), then the more cost operating units (called the intermediate units), followed by the costliest operating units (called the peaker units), and so on. This means that the least cost operating units occupy the lower levels in the LDC, and the most expensive operating units occupy the upper levels in the LDC.

4.2 Method of evaluation of the expected energy supplied

The expected energy supplied (ϵES) by each unit available and being operated in the system can be evaluated by using the above concept of the expected energy not supplied (ϵENS), as shown below:

$$\epsilon ES_i = \epsilon ENS_{i-1} - \epsilon ENS_i \text{ MWh/year} \tag{9}$$

This method adopts a priority loading order, i.e., the generating units are loaded according to their least operating costs. The procedure applied is described above (see **Figure 5**).

The process of the above figure can be interpreted in the following steps:

- The load duration curve is implemented, as it is the type of curve that is widely used in power system reliability evaluation and planning for its convenience and flexibility. It is derived from the ordinary load curve and hence can be defined as "the arrangement of all load levels in a descending order of magnitude."
- The expected energy not supplied (ϵENS_0) before any unit is operated is the total area under the LDC.

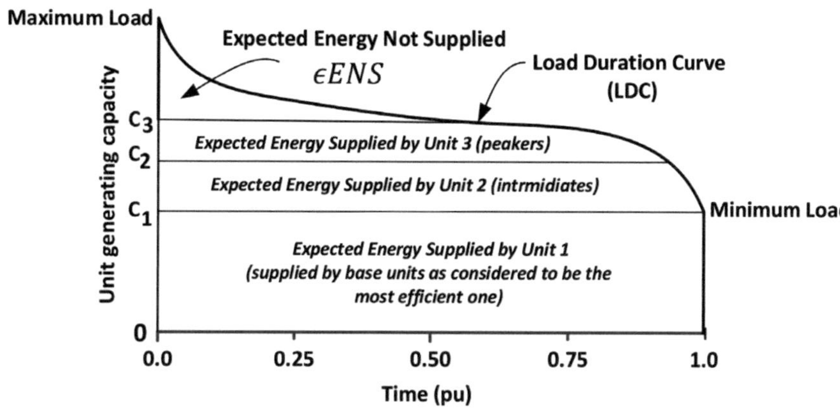

Figure 5.
Load duration curve displaying units loading priority.

- When the first unit (C_1) is loaded according to the priority loading level #1, it will occupy the area ($0 - C_1$) and shifts the new expected energy not supplied (ϵENS_1) upward (i.e., above C_1). Therefore, the expected energy supplied by unit C_1 (ϵES_1) will be $\epsilon ES_1 = (\epsilon ENS_0) - \epsilon ENS_1$.

- When the second unit (C_2) is loaded according to the priority loading level #2, it will occupy the area ($C_1 - C_2$) and then shift the new expected energy not supplied (ϵENS_2) upward above C_2. Therefore, the expected energy supplied by unit C_2 (ϵES_2) will be $\epsilon ES_2 = \epsilon ENS_1 - \epsilon ENS_2$.

- When the third unit (C_3) is operated according to the priority loading level #3, it will occupy the area $C_2 - C_3$ and then shift expected energy not supplied (ϵENS_3) above C_3, and then the process ends, and the remaining expected energy not supplied will be above C_3. As such, the expected energy supplied by unit C_3 (ϵES_3) will be $\epsilon ES_3 = \epsilon ENS_2 - \epsilon ENS_3$.

The following example shows an industrial compound case having two generating units, namely, 80 MW and 60 MW, which are assigned with a loading priority of "1" and "2," respectively. The expected energy supplied (**ϵES**) and the energy index of reliability (EIR) are both to be determined, so as to optimize its energy production with least possible operating cost.

Example: A power plant has the following data:

Capacity (MW)	FOR	Loading priority
80	0.06	1
60	0.03	2

The LDC is to be considered as a straight line connecting a maximum load of 160 MW and a minimum load of 80 MW (**Figure 6**). If the total operating time is 100 hours, evaluate the following:

a. The expected energy supplied (**ϵES**)by each unit in the system

b. The energy index of reliability (**EIR**) of the system

The solution hereto is to, first, calculate the expected energy not supplied before any unit in the system is being loaded (ϵENS_0), i.e., at 0 MW, which is

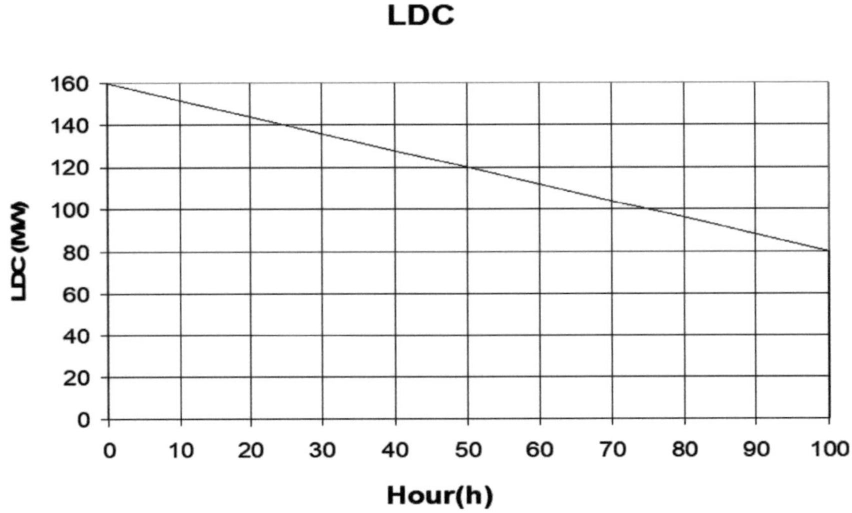

Figure 6.
Load duration curve for the given example.

$$\epsilon ENS_0[0\ \mathrm{W}] = 80 \cdot 100 + \tfrac{1}{2} \cdot 100 \cdot (160 - 80) = 12000\,\mathrm{MWh}\ \text{(Area under the LDC)}$$
$$\epsilon ENS_0[0\ \mathrm{W}] = 12000\ \cdot 1 = 12000\ \ \mathrm{MWh}$$

Now start loading the units starting with the first unit (i.e., 80 MW as unit no. 1 for the priority order no. 1). This is shown in **Table 1**.

$$\epsilon ENS_1\,[0\,\mathrm{MW}] = 12000\ \cdot 0.06 = 720\ \mathrm{MWh}$$

$$\epsilon ENS_1[80\,\mathrm{MW}] = \left[\frac{1}{2} \cdot 100 \cdot (160 - 80)\right](0 \cdot 94) = 3760\,\mathrm{MWh}$$

Therefore, the expected total energy not supplied after the first unit is being added ($\epsilon TENS_1$) will be

$$\epsilon TENS_1\ = 720 + 3760 = 4480\ \ \mathrm{MWh}$$

Therefore, the expected energy supplied by the unit 80 MW (ϵES_1) can be evaluated as

$$\epsilon ES_1 = \epsilon \mathrm{MW}\, S_0 - \epsilon TENS_1 = 12000\ \text{-}\mathrm{MWh}\ = \mathbf{7520}\ \ \mathrm{MWh}$$

Now, loading the second unit (i.e., unit of 60 MW as unit no. 2 for the priority order no. 2), the new CAT in **Table 2** will be

$$\epsilon ENS_2[0\ \mathrm{MW}]\quad = 1200 \times 0.0018 = 21.6\quad \mathrm{MWh}$$

$$\epsilon ENS_2[60\,\mathrm{MW}]\ = 6000 \times 0.0582 = 349.2\ \ \mathrm{MWh}$$

$$\epsilon ENS_2[80\,\mathrm{MW}]\ = 4000 \times 0.0282 = 112.8\ \ \mathrm{MWh}$$

$$\epsilon ENS_2[140\,\mathrm{MW}] = 250 \times 0.9118 = 227.95\ \ \mathrm{MWh}$$

System capacity (MW)	Availability
0	0.06
80	0.94

Table 1.
System CAT at priority order level no. 1.

System capacity (MW)	Availability
0	0.06 × 0.03 = 0.0018
60	0.06 × 0.97 = 0.0582
80	0.94 × 0.03 = 0.0282
140	0.94 × 0.97 = 0.9118

Table 2.
System CAT at priority order level no. 2.

Therefore, the expected total energy not supplied after the second unit is being added ($\epsilon TENS_2$) will be

$$\epsilon TENS_2 = 21.6 + 349.2 + 112.8 + 227.95 = 711.55 \text{ MWh}$$

As such, the expected energy supplied by the unit 80 MW (ϵES_2) can be evaluated as

$$\epsilon ES_2 = \epsilon TENS_1 - \epsilon TENS_2 = 4480 - 711.55 = \mathbf{3608.45} \text{ MWh}$$

Hence, unit no. 1 (80 MW) will serve 7520 MWh, and unit no. 2 (60 MW) will serve 3608.45 MWh.

Now, the final remaining expected total energy not supplied ($\epsilon TENS$) MWh for this system is 711.55 MWh, and the system energy index of reliability (EIR) can be evaluated as

$$EIR = 1 - \frac{\epsilon TENS_2}{\epsilon TENS_0} = 1 - \frac{711.55}{12000} = 0.94$$

5. Applications of reliability indices in power system planning

Optimal reliability evaluation is an essential step in power system planning processes in order to ensure dependable and continuous energy flow at reasonable costs. Therefore, the reliability index, namely, the loss of load expectation ($LOLE$), discussed in Section 3.4 along with the other complementary indices discussed in Sections 3.5–3.7 can be quite useful. Indeed, in order to substantiate and verify the applicability thereof, these indices have been applied to a real power system case study situated in the northern part of the Kingdom of Saudi Arabia. This power system is supposed to serve a major populated community with a potential future commercial and industrial load growth acknowledging the Kingdom's "Vision 2030."

The various reliability and economic models incorporated in the planning process are portrayed in **Figure** 7 and can be summarized as follows:

1. DATMOD: data model retrieving and organizing all studied system needed data like load duration curve (LDC), capacity outrage probability table

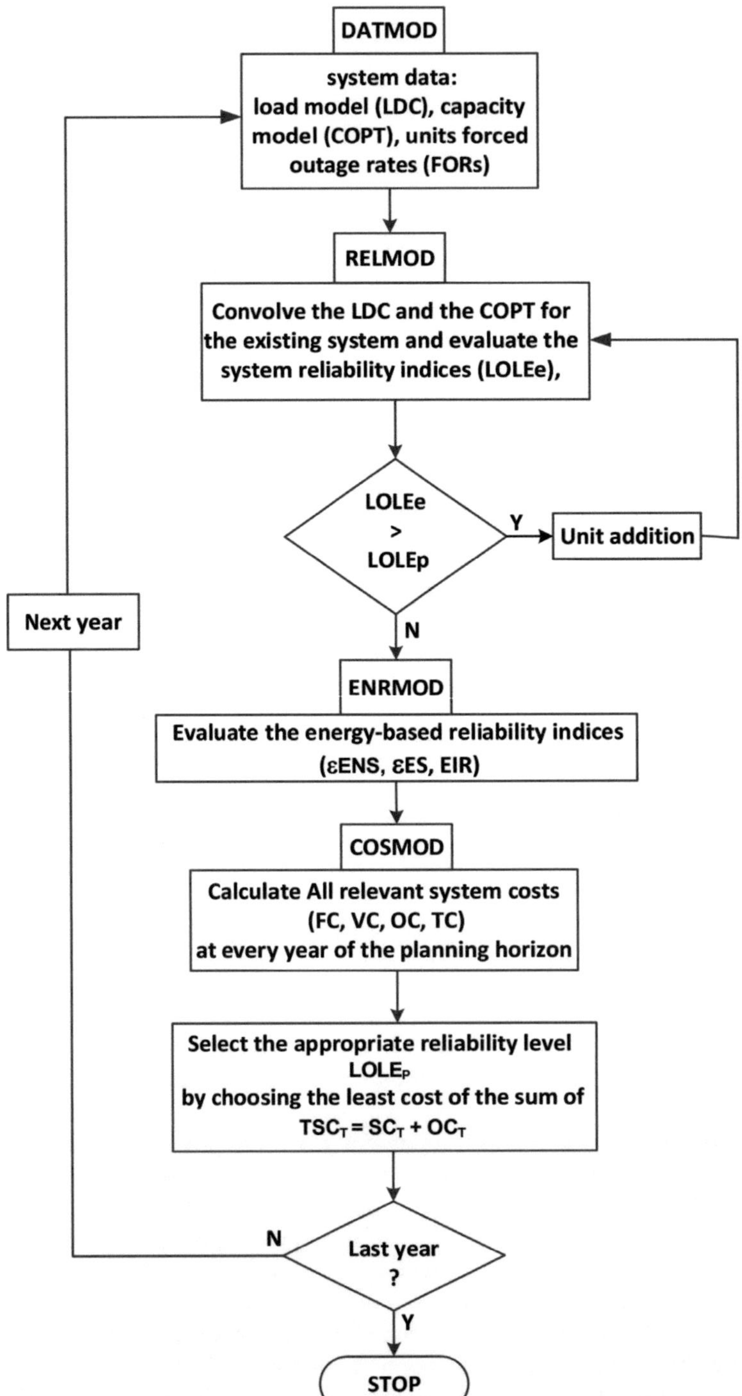

Figure 7.
Planning process for optimal reliability levels.

(COPT), and forced outage rates (FORs) pertinent to all generating units either residing in the system or newly added unit(s)

2. RELMOD: reliability model that evaluates studied system reliability ($LOLE$) levels at every year of the planning period and decides whether a unit(s) is needed to be added or to be postponed until it is required

3. ENRMOD: energy model which assesses expected energy supplied (ϵES) by the generating units residing in or added to the system and also estimates the remaining expected energy not supplied (ϵENS) and the energy reliability index (EIR)

4. COSMOD: cost model that estimates all cost pertinent to the system (system cost, outage cost, total cost) to be compared and assessed for optimum use

In order to obtain the most appropriate range of reliability levels, the system cost should be weighted with the estimated outage cost. System costs include fixed cost in terms of unit installation cost and variable cost in terms of fuel and maintenance cost. The outage cost (OC) forms a major part in the total system cost. These costs are associated with the demanded energy but cannot be supplied by the system due to severe outages occurrences, and is known as the expected energy not supplied, (ϵENS).

Outage cost is usually borne by the utility and its customers. The system outage cost includes loss of revenue, loss of goodwill, loss of future sales, and increased maintenance and repair expenditure. However, the utility losses are seen to be insignificant compared with the losses incurred by the customers when power interruptions and energy cease occur. The customers perceive power outages and energy shortages differently according to their categories. A residential consumer may suffer a great deal of anxiety and inconvenience if an outage occurs during a hot summer day or deprives him from domestic activities and causes food spoilage. For a commercial user, he/she may also suffer a great hardship and loss of being forced to close until power is restored. Also, an outage may cause a great damage to an industrial customer since it disrupts production and hinders deliveries.

The overall system cost depicts the overall cost endured by the customers as a value of uninterrupted power flow. The outcome of the process yields the results

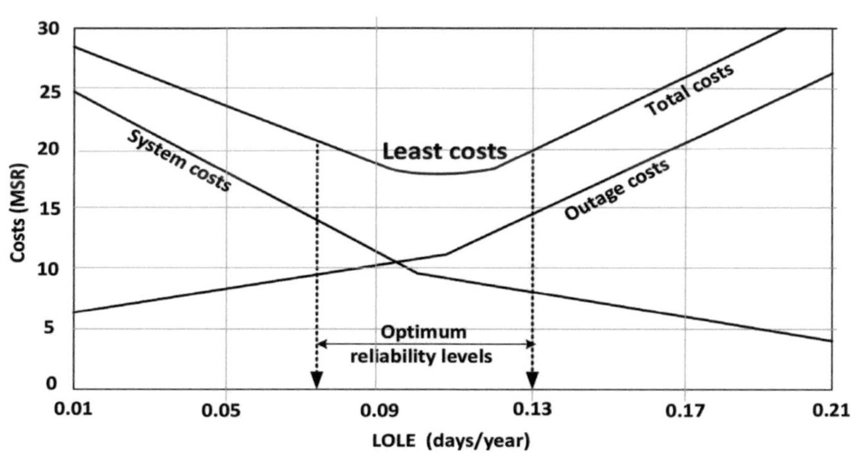

Figure 8.
Variations of LOLE levels with costs.

shown by **Figure 8**, in which system cost (*SC*) increases as the reliability level increases. At the same time, the outage cost (*OC*) decreases because of reliability improvement and adequate generating capacity additions. The most optimal reliability levels vary between 0.07 and 0.13 days/year (see **Figure 8**). However, in some cases adding new capacity may not signify the ideal solution to meet increasing future loads and maintain enhanced reliability levels. Therefore, it is better to improve an operating unit's performance through regular preventive maintenance. Likewise, establishing a good cooperation between the supply side (electric company) and the demand side (the customers) through well-coordinated load management strategies may further improve financial performance (1£ = 4.5 SR).

6. Applications of reliability indices in power system interconnection

6.1 Introduction

A review of the main advantages of electrical interconnection between electrical power systems is summarized as follows:

- When connecting isolated electrical systems, each system needs a lower generation reserve than the reserve when it is isolated and at a better level of reliability.

- When interconnecting isolated electrical systems, it is possible to share the available reserve so that each system maintains a lower level of reserve before being interconnected. This will result in both lower installation costs (fixed costs) and decreased operation costs (variable costs).

- The electrical connection reduces the fixed and operating costs of the total installed capacity.

- In emergency and forced outage conditions, such as breakdowns, multiple interruptions, and the simultaneous discharge of several generators, which may cause a capacity deficit that is incapable of coping with current loads and possibly a total breakdown of the electrical system as a whole, electrical interconnection helps to restore the state of stability and reliability of electrical systems.

- The interconnection of power systems enables the exchange of electrical energy in a more economical manner, as well as the exchange of temporal energy and the utilization of the temporal variation in energy demand.

- The electrical connection through the construction of larger power plants with higher economic return and reliability increases the degree of cooperation and the sharing of potential opportunities and possibilities that are available between the electrical systems.

- By nature, the various loads do not have peak values at the same time. As a result of this variation in peak loads (maximum demands), the load of the interconnected systems is less than the total load of each system separately, thus reducing and saving the total power reserve for systems.

6.2 Method of implementation

The above brief review of the main advantages and merits of electrical interconnection from an economic and technical point of view highlights the usefulness and importance of conducting electrical interconnection studies between the systems as they relate to the cost of capital and operational costs on the one hand and the improvement of their levels and performance on the other. Such studies are especially significant after the completion of the infrastructure of electrical systems. Indeed, the next step is to seriously consider linking electrical systems through unified national networks throughout the widespread Kingdom.

Most power systems have interconnections with neighboring systems. The interconnection reduces the amount of generating capacity required to be installed as compared with that which would be required without the interconnection. The amount of such reduction depends on the amount of assistance that a system can get, the transfer capability of the tie-line, and the availability of excess capacity reserve in the assisting systems.

One objective to be mentioned in this context is to evaluate the reliability benefits associated with the interconnection of electric power systems. Therefore, this study is focused on the reliability evaluation of two systems that may be viewed upon as both isolated systems and as interconnected systems. The analysis of this type explores the benefits that may accrue from interconnecting systems rather than being isolated as well as deciding viable generation expansion plans.

A 5-year expansion plan for systems A and B assuming a reliability criterion of 0.1 days/year (0.1–0.6 frequently quoted as appropriate values in most industrial countries) was determined. The analysis represents the expansion plans for both systems as being isolated and interconnected. An outcome of these expansion plans is shown in **Figure 9**.

If the two systems (A and B) are reinforced whenever the reliability index (risk level) falls below the prescribed level (i.e., $LOLE_p = 0.1\ \text{d/y}$) at any year of the planning horizon, the results shown in the following table exhibits that the number

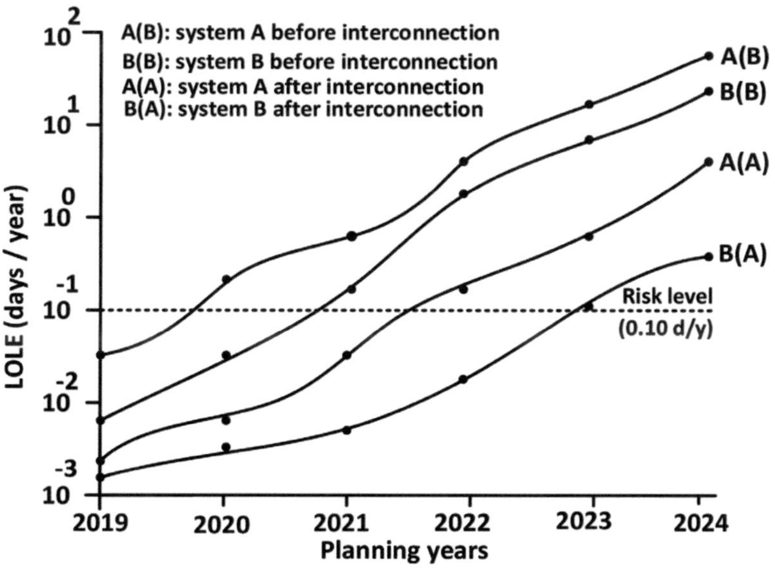

Figure 9.
LOLE levels before and after systems interconnection.

of added units and their cost are reduced if the two system are interconnected rather than being isolated.

System costs as isolated and interconnected:

System	Isolated			Interconnected		
	No. of unit	Cost (MSR)	*eENS* (MWh)	No. of unit	Cost (MSR)	*eENS* (MWh)
A	4	12.63	5.652	2	9.44	1.054
B	2	16.42	4.852	1	8.75	2.045

Therefore, it can be concluded from the above analysis that both systems will benefit from the interconnection. The reliability of both systems can be improved, and consequently the cost of service will be reduced through interconnection and reserve sharing. However, this is not the overall saving because the systems must be linked together in order to create an integrated system. The next stage must, therefore, assess the economic worth that may result from either interconnection or increasing generating capacity individually and independently.

7. Transmission and distribution reliability evaluation

7.1 Introduction

Since embarking on the national industrial development and the industrials program in the Kingdom of Saudi Arabia, the Ministry of Energy, Industry and Mineral Resources launched two solar PV projects with a combined generation capacity of 1.51 GW enough to power 226,500 households. These projects will be tendered by mid-2019 to attract a total investment of $1.51 billion Saudi Riyals creating over 4500 jobs during construction, operations, and maintenance [13]. The program will be phased and rolled out in a systematic and transparent way to ensure that the Kingdom benefits from the cost-competitive nature of renewable energy. The National Renewable Energy Program aims to substantially increase the share of renewable energy in the total energy mix, targeting the generation of 27.3 gigawatts (GW) of renewable energy by 2024 and 58.7 GW by 2030. This initiative sets out an organized and specific road map to diversify local energy sources, stimulate economic development, and provide sustainable economic stability to the Kingdom in light of the goals set for Vision 2030, which include establishing the renewable energy industry and supporting the advancement of this promising sector.

7.2 Role of the government in the electricity sector

As a result of the continuous subsidy and generous support of the government for the electricity sector, the ministry has been able to accomplish many electrical projects in both urban and rural areas, resulting in electric services that can reach remote areas and sparsely populated areas, over rough roads and rugged terrain. In fact, electric services require large sums of money to finance, build, operate, safeguard, and sustain. Another important component that must be considered along with the continuous operation and maintenance expenditures is fuel costs. Therefore, constant maintenance measures ought to be implemented to ensure the level and continuity of the flow of electrical energy without fluctuation, decline, or interruption.

The expansion of the electricity sector during the last three decades has resulted in the many electricity companies throughout the Kingdom being integrated into what was known, for a short time, as "the Saudi Consolidated Electric Companies (SCECOs)." These companies later merged into a single more reliable, efficient, and less expensive company known as the "Saudi Electricity Company (SEC)." Moreover, some areas (Eastern and Central) have been linked via a tie-line in order to prepare for the integration of the entire Kingdom under a unified national network.

Experts and planners of electrical power systems find it economically and technically unfeasible to increase the electrical capabilities of electric power plants that are often isolated, dispersed, and distant. However, after the completion of the structures of these systems, the next and natural step, to achieve advantages and benefits, is to connect these electric power systems to each other through unified transmission networks. Undoubtedly, linking these power systems will both reduce the cost of construction and provide reserve and fuel, all while increasing the strength of the electrical system and maximizing its capability to meet current and future electric loads.

7.3 Practical example

One practical example demonstrating the evolving of industry of electric sector in the Kingdom of Saudi Arabia will be shown in this section. The availability of network can be analyzed in a similar manner to that used in generating capacity evaluation (Section 3.1). Therefore, the probability of failing to satisfy the criterion of service adequacy and continuity can be evaluated. Provided the appropriate component reliability indices are known, it is relatively simple to evaluate the expected failure rate (λ) of the system, the average duration of the outage I, and the unavailability or annual outage time (U). To do this, the values of λ, r, and U are required for each component of the system.

7.3.1 State probabilities

The state-space transition diagram for a two-component system is shown in **Figure 10**.

The probability of a component being in the up state is $\frac{\mu}{\lambda+\mu}$.

Also, the probability of a component being in the down state is $\frac{\lambda}{\lambda+\mu}$.

$$\text{Probability of being in state 1} = \frac{\mu_1}{\mu_1+\lambda_1}\cdot\frac{\mu_2}{\mu_2+\lambda_2}$$

$$\text{Probability of being in state 2} = \frac{\lambda_1}{\mu_1+\lambda_1}\cdot\frac{\mu_2}{\mu_2+\lambda_2}$$

$$\text{Probability of being in state 3} = \frac{\mu_1}{\mu_1+\lambda_1}\cdot\frac{\lambda_2}{\mu_2+\lambda_2}$$

$$\text{Probability of being in state 4} = \frac{\lambda_1}{\mu_1+\lambda_1}\cdot\frac{\lambda_2}{\mu_2+\lambda_2} \tag{10}$$

The most accurate method for analyzing networks including weather states is to use the Markov modeling. However, this becomes impractical for all except the simplest system. Instead, therefore, an approximate method is used based upon simple rules of probability.

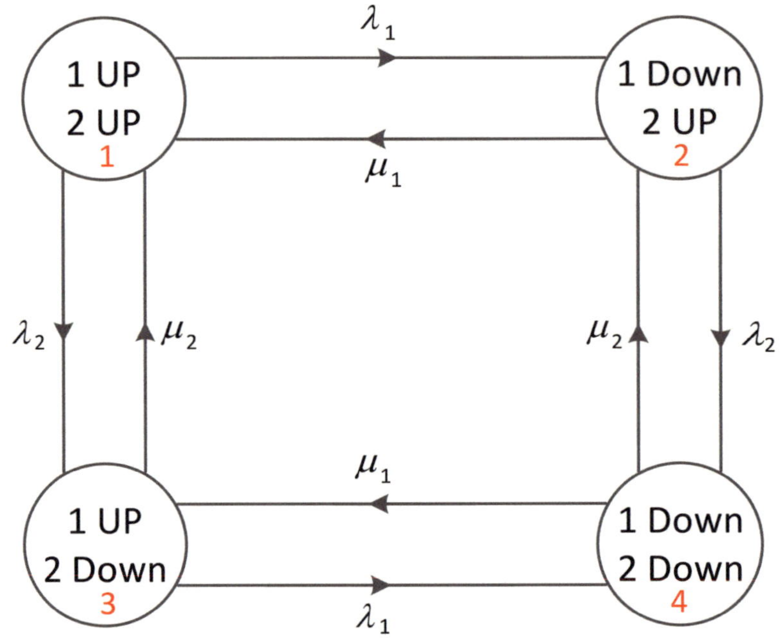

Figure 10.
State-space diagram for two-component system, where λ is the failure rate and μ is the repair rate $= \frac{1}{r}$ (r = repair time).

7.3.2 Series components

The requirement is to find the reliability indices of a single component that is equivalent to a set of series-connected components as shown in **Figure 11.**

If the components are in series from a reliability point of view, both must operate, i.e., be in upstate, for the system to be successful, i.e., the upstate of a series system is state 1 of the state-space diagram shown in **Figure 11**.

From the above equation (state 1), the probability of being in this upstate is

In addition, since $\mu = \frac{1}{r}$, the above equation becomes

$$r_s = \frac{\lambda_1 r_1 + \lambda_2 r_2 + \lambda_1 \lambda_2 r_1 r_2}{\lambda_s} \tag{11}$$

$$= \frac{\lambda_1 r_1 + \lambda_2 r_2}{\lambda_s} \tag{12}$$

$$= \frac{\sum_{i=1}^{n} \lambda_i \, r_i}{\lambda_s} \tag{13}$$

Also, the rate of transition from state 1 of the two-component state-space diagram is $\lambda_1 + \lambda_2$, therefore

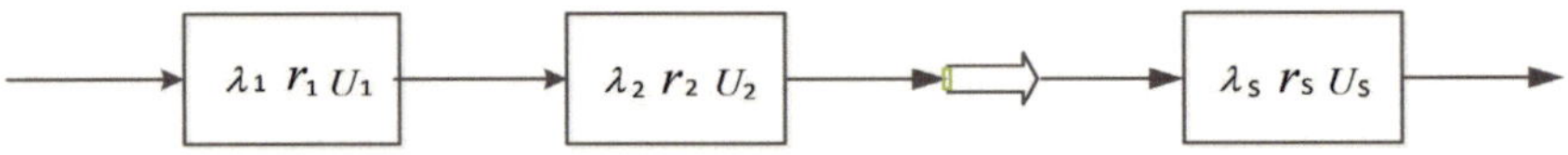

Figure 11.
State-space diagram for two-component system.

$$\lambda_s = \lambda_1 + \lambda_2 = \sum_{i=1}^{n} \lambda_i \tag{14}$$

$$r_s = \frac{\sum_{i=1}^{n} \lambda_i r_i}{\sum_{i=1}^{n} \lambda_i} \tag{15}$$

Thus, the unavailability for series systems (U_S) can be expressed as

$$U_S = \lambda_s r_s \tag{16}$$

$$= \sum \lambda r \tag{17}$$

In particular, the order of evaluation is usually $\lambda_s (= \sum \lambda)$, $U_s (= \sum \lambda r)$ and $r_s (= U_s / \lambda_s)$.

Although these equations were derived from the assumption of exponential distribution, they are expected or average values and can be shown to be valid irrespective of the distribution assumption.

7.3.3 Parallel components

Many systems consist of both series and parallel connections. These systems can be seen in transmission lines, in combinations of transformers, cables, feeders, relays, protection and control devices, etc. As an example, **Figure 12** displays two parallel lines that are both connected in series with another line. In these situations, and from a reliability point of view, it is essential to consequently reduce the network in order to estimate its overall reliability. This is accomplished by repeatedly combining sets of parallel and series components into equivalent network components until a single component remains. The reliability of the last component is equal to the reliability of the original system (**Figure 12**).

In this case, the requirement is to find the indices of a single component that is equivalent to two parallel components as shown in **Figure 12.**

If the components are in parallel from a reliability point of view, both must fail for resulting in a system failure, i.e., the down state of a parallel system is state 4 of the state-space diagram shown in **Figure 10**. From (10), the probability of being in this downstate is

$$\frac{\lambda_1 \lambda_2}{(\mu_1 + \lambda_1)(\mu_2 + \lambda_2)} = \frac{\lambda_p}{\lambda_p + \mu_p} \tag{18}$$

Also, the rate of transition from state 4 of the two-component state-space diagram is $\mu_1 + \mu_2$.

Therefore $\mu_p = \mu_1 + \mu_2$

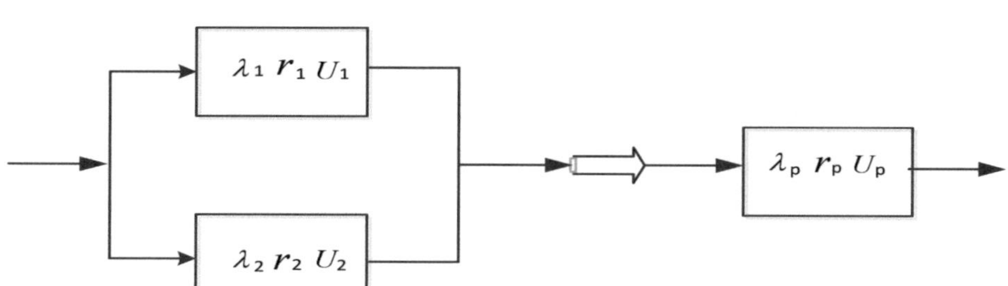

Figure 12.
State-space diagram for a two-component system.

$$\frac{1}{r_p} = \frac{1}{r_1} + \frac{1}{r_2} \tag{19}$$

or

$$r_p = \frac{r_1 r_2}{r_1 + r_2} \tag{20}$$

From the above equations, it yields that

$$\lambda_p = \frac{\lambda_1 \lambda_2 (r_1 + r_2)}{1 + \lambda_1 r_1 + \lambda_2 r_2} \tag{21}$$

$$= \lambda_1 \lambda_2 (r_1 + r_2) \tag{22}$$

Thus, the unavailability for parallel systems (U_p) can be expressed as

$$U_p = \lambda_p r_p \tag{23}$$

In practice, the order of evaluation is usually

$$\lambda_s = \sum \lambda \qquad U_s = \sum \lambda r \qquad \text{and} \qquad r_s = \frac{U_s}{\lambda_s}$$

Although these equations were derived from the assumption of exponential distribution, they are expected or average values and can be shown to be valid irrespective of the distributional assumption.

Example (series/parallel): To illustrate the applications of these techniques, let us consider the transmission lines supplying the newly large industrial park constructed near Riyadh city (the capital of the KSA) within what is called "industrial cities" in the main cities of the KSA. The transmission lines with their data load points are given below (see **Figure 13**). It is required to evaluate the load point (busbar) reliability indices at busbars B and C.

To find the indices at busbar B, lines 1 and 2 must be combined in parallel using Eq. (22):

$$\lambda_B = \lambda_1 \lambda_2 (r_1 + r_2)$$
$$= 0.5 \times 0.5(5 + 5)/8760$$
$$= 2.854 \times 10^{-4} f/y$$

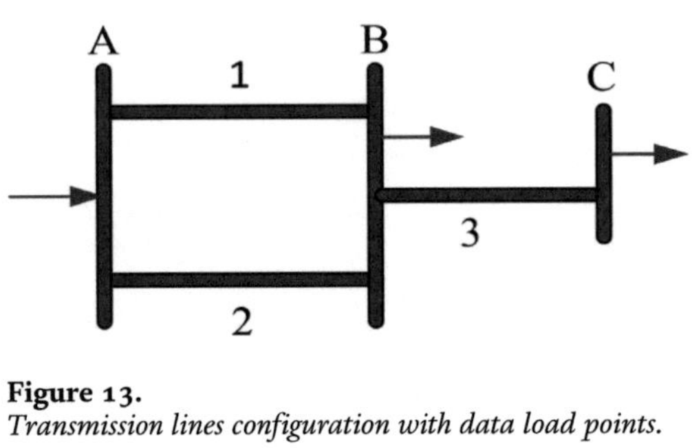

Line #	(f / yr)	r (hrs.)
1	0.5	5
2	0.5	5
3	0.1	10

Figure 13.
Transmission lines configuration with data load points.

where 8760 is the total number of hours in a year, using Eq. (20)

$$\begin{aligned} r_B &= \frac{r_1 r_2}{r_1 + r_2} = \frac{5 \times 5}{5 + 5} = 2.5 \text{ h.} \\ U_B &= \lambda_B r_B \\ &= 2.854 \times 10^{-4} \times 2.5/8760 \\ &= 8.145 \times 10^{-8} \text{ yrs/yrs} = \textit{probability} \\ &= 7.135 \times 10^{-4} \text{ h/yrs.} \end{aligned}$$

To find indices at busbar C, lines 1 and 2 must be combined in parallel (as done above) and then combined with line 3 in series, using Eq. (14):

$$\begin{aligned} \lambda_C &= \lambda_B + \lambda_3 \\ &= 2.854 \times 10^{-4} + 0.1 \\ &= 1.003 \times 10^{-1} f/yr \\ r_C &= \frac{U_B + \lambda_3 r_3}{\lambda_C} \\ r_C &= \frac{7.135 \times 10^{-4} + 0.1 \times 10}{1.003 \times 10^{-1}} \\ &= 9.977 \text{ h.} \end{aligned}$$

Using Eq. (23)

$$\begin{aligned} U_C &= \lambda_C r_C \\ \therefore U_C &= 1.003 \times 10^{-1} \times 9.977 \\ &= 1.001 \text{ h/yrs.} \end{aligned}$$

In this case, it is seen that the indices of busbar C are dominated by the indices of line 3. This is clearly expected since busbar C will be lost if either line 3 or lines 1 and 2 simultaneously fail. Consequently, loss of line 3 is a first-order event, and loss of lines 1 and 2 are a second-order event. It must be stressed that this is only true if the reliability indices of the components are comparable; if the component forming the low-order event is very reliable and the components forming the higher order events are very unreliable, the opposite effect may occur.

7.3.4 Network reduction for failure mode analysis

In some cases, some critical or unreliable areas become absorbed into equivalent elements and become impossible to identify. The alternative is to impact the system and compose a list of failure nodes, i.e., component outages that must overlap to cause a system outage. These overlapping outages are effectively parallel elements and can be combined using the equations for parallel components. Any one of these overlapping outages will cause system failure and therefore, from a reliability point of view, are effectively in series. The system indices can therefore be evaluated by applying the previous series equations to these overlapping outages.

The following case study showcases the existing tie-line interconnecting the eastern region (ER) with the central region (CR) (400 km apart) in the Kingdom of Saudi Arabia (KSA). The ER is actually the incubator of the oil industry and all its refineries and infrastructures. Riyadh is located in the CR, which is the domicile of the Saudi Electric Company (SEC). The latter is envisioning tremendous expansion with vast increasing industrial future loads. Therefore, a huge bulk of electric power is transferred from the ER to the CR via the interconnecting tie-line. Therefore, to evaluate its reliability using the concepts and methodology stated above, the tie-line (see **Figure 14**) is considered bearing the following data:

a. Using network reduction

Combing elements 1 and 3 in series as in Eq. (12) gives:

$$r_{1,3} = \frac{\lambda_1 r_1 + \lambda_3 r_3}{\lambda_s} = \frac{0.5 \times 10 + 0.01 \times 100}{0.51} = 11.76 \ \ \mathbf{hr}$$

The indices of components 2 and 4 combined will be identical:

$$r_{2,4} = \frac{\lambda_2 r_2 + \lambda_4 r_4}{\lambda_s} = \frac{0.5 \times 10 + 0.01 \times 100}{0.51} = 11.76 \ \ \mathbf{hr}$$

The indices for the load point are

$$\lambda_s = 6.984 \times 10^{-4} \ f/y \quad r_2 = 5.88 \ \mathbf{hr}$$

$$U_s = 4.1066 \times 10^{-3} \quad \mathbf{hr/yr}$$

b. Using failure modes analysis

Overlapping outages	**λ (*f/yr*)**	***r* (h)**	***U* h (h/yr)**
1 and 2	5.7080×10^{-14}	5	2.854×10^{-14}
1 and 4	0.6279×10^{-14}	9.091	5.708×10^{-14}
2 and 3	0.6279×10^{-14}	9.091	5.708×10^{-14}
3 and 4	0.0228×10^{-14}	50	1.142×10^{-14}
	$6.987 \times 10^{-14} = \lambda_s$	$5.88 = r_s$	4.110×10^{-14} $\lambda_s \times r_s$

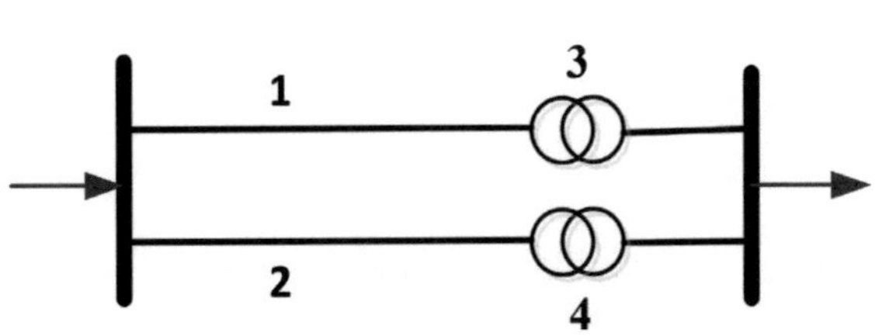

Component	λ (f / yr)	r (hrs.)
1	0.5	10
2	0.5	10
3	0.01	100
4	0.01	100

Figure 14.
The tie-lines configuration with data load points.

Although the second method seems longer, it is worth noting that it gives a greater deal of information. It indicates that the failure rate and unavailability are mainly due to the overlapping failures of the two lines; however, the average outage duration is mainly due to the overlapping outages of the two transformers. This information, which is vital in assessing critical areas and indicating the areas requiring more investment, is not given by the network reduction technique.

8. Customer-based reliability indices

The most widely used reliability indices are averages that weight each customer equally. Customer-based indices are popular with electric companies [14] since a small residential customer has just as much importance as a large industrial customer. Regardless of the limitations they have, these are generally considered acceptable techniques showing adequate measures of reliability. Indeed, they are often used as reliability benchmarks and improvement targets. The formulae for customer-based indices include:

8.1 System average interruption frequency index (SAIFI)

SAIFI is a measure of how many sustained interruptions an average customer will experience over the course of a year. This measure can be defined as

$$\text{SAIFI} = \frac{\text{Total number of customers interruptions}}{\text{Total number of customers served}}\ (\text{inter/cust}) \tag{24}$$

For a fixed number of customers, the only way to improve SAIFI is to reduce the number of sustained interruptions experienced by customers.

8.2 System average interruption duration index (SAIDI)

SAIDI is a measure of how many interruption hours an average customer will experience over the course of a year. For a fixed number of customers, SAIDI can be improved by reducing the number of interruptions or by reducing the duration of these interruptions. Since both of these reflect reliability improvements, a reduction in SAIDI indicates an improvement in reliability. This measure can be defined as

$$S\text{AIDI} = \frac{\text{Total customers interruptions durations}}{\text{Total number of customers served}}\ (\text{h/cust}) \tag{25}$$

8.3 Customer average interruption duration index (CAIDI)

CAIDI is a measure of how long an average interruption lasts and is used as a measure of utility response time to the system contingencies. CAIDI can be improved by reducing the length of interruptions but can also be reduced by increasing the number of short interruptions. Consequently, a reduction in CAIDI does not necessarily reflect an improvement in system reliability. This measure can be defined as

$$\text{CAIDI} = \frac{\text{Total customers interruptions durations}}{\text{Total number of customers interruptions}}\ (\text{h/cust}) \tag{26}$$

8.4 Average service availability index (ASAI)

ASAI is the customer-weighted availability of the system and provides the same information as SAIDI. Higher ASAI values reflect higher levels of system reliability. This measure can be defined as

$$\text{ASAI} = \frac{\text{Customer hours service availability}}{\text{Customer hours service demand}} \text{ (pu)} \tag{27}$$

9. Conclusions

This chapter consists of eight sections that can be briefly summarized as follows:

Section 1 starts with an introduction that indicates the importance and viable role of reliability evaluation in power system planning with selected relevant references to its nature subject matter.

Section 2 discusses the types of equipment outages, particularly the severe ones that may cause the machine(s) to be out of service unexpectedly in critical conditions that can compromise the ability of the system to supply the load.

Section 3 reviews some basic theories, assumptions, and mathematical expressions for the reliability evaluation such as the well-known "loss of load expectation" index and with other important complementary reliability indices.

Section 4 exhibits a new computation method for the energy produced by each generating unit loaded to the system.

Section 5 demonstrates how the reliability indices can be of significant tools in assessing system planners to arrive at the most appropriate reliability levels that can assure both continuous supply as well as maintaining the least operating cost.

Section 6 highlights the main merits and advantages of electrical interconnection among dispersed and isolated power systems from an economic and reliability point of view.

Section 7 shows the application of the frequency and duration (F&D) indices used in reliability evaluation of transmission lines and distribution networks. These indices are implemented in some industrial zones in a fast-developing country in accordance with its envisaged 2030 vision.

Section 8 reveals the most widely used customer-based reliability indices by most of the electric companies since the residential sector has just as much importance as the industrial sector. These indices show adequate measures of reliability benchmarks and improvement targets.

Appendix A. Power system costs

There are several costs that are associated with power system planning and can be manifested in the following sections.

A.1 Fixed cost

The fixed cost (FC) represents the cash flow at any stage of the planning horizon resulting from the costs of installing new generating units during the planning period. It depends on the current financial status of the utility, the type and size of generating units, and the cost of time on money invested during the planning period. The total fixed costs (FC_T) for unit(s) being installed can be computed as

$$FC_T = \sum_t \sum_k (CAP_k \cdot CC_k \cdot NU_k)^t \tag{A.1}$$

where
CAP_k: *unit capacity added to the system of type k*; -
CC_k : *capital cost of unit of type k (SR/kW)*;
NU_k: *number of unit(s) added to the system of type k at each interval of time t.*;
t: *interval period of time considered in the planning horiozon,* $t = 1 \ldots T$.

A.2 Variable cost

The variable cost (VC) represents the cost of energy supplied by the system. It is affected by the load variation, the type and size of generating units, and the number of hours of operation. Also, these costs are related to the cost of operation and maintenance (fuel, scheduled maintenance, interim spare parts, staffing, wages, and miscellaneous expenses) and can be evaluated as

$$VC_T = \sum_t \sum_k (\epsilon ES_k \cdot ESC_k \cdot NU_k)^t \tag{A.2}$$

where ϵES_k: expected energy supplied by unit of type k; ESC_k: energy supplied cost of unit of type k (SR/kWh).

The total system costs (SC_T) for the entire expansion plan can be estimated by summing all the above individual costs at every stage of the planning period as being expressed in the following equation:

$$SC_T = FC_T + VC_T \tag{A.3}$$

A.3 Outage cost

The outage costs, i.e., the cost of the expected energy not supplied (ϵENS), were previously presented and discussed in Section 5. One method of evaluating ϵENS is described in [8]. Therefore, estimating the outage cost (OC) is to multiply the value of that ϵENS by an appropriate outage cost rate (OCR), as follows:

$$OC_T = \sum_t (\epsilon ENS \cdot OCR)^t \tag{A.4}$$

where ϵENS is the expected energy not supplied (kWh lost) and OCR is the outage cost rate in SR/kWh.

The overall cost of supplying the electric energy to the consumers is the sum of system cost that will generally increase as consumers are provided with higher reliability and customer outage cost that will, however, decrease as system reliability increases or vice versa. This overall system cost (OSC) can be expressed as in the following equation:

$$OSC_T = SC_T + OC_T \tag{A.5}$$

The prominent role of outage cost estimation, as revealed in the above equation, is to assess the worth of power system reliability by comparing this cost (OC) with the size of system investment (SC) in order to arrive at the least overall system cost that will establish the most appropriate system reliability level that ensures energy continuous flow as well as the least cost of its production.

As witnessed in **Figure 8**, the incorporation of customer outage costs in investment models for power system expansion plans is very difficult for planners in fast-developing countries. This difficulty stems principally either from the lack of system records of outage data, failure rate, frequency, duration of repair, etc. or the failure to carry out customer surveys to estimate the impact and severity of such outages in terms of monetary value.

Author details

Abdullah M. Al-Shaalan
King Saud University, Riyadh, Saudi Arabia

*Address all correspondence to: shaalan@ksu.edu.sa

References

[1] Billinton R, Allan RN. Reliability Evaluation of Power Systems. London: Pitman Advanced Publishing Program; 1984

[2] Billinton R. Reliability Assessment of Bulk Electric Systems. Publications No. 148. Canada: Power System Research Group, University of Saskatchewan; 2007. p. 75

[3] Endrenyi J. Reliability Modelling in Electric Power Systems. Wiley International Publication; 1978

[4] Grigsby LL. "Power Systems", Power System Planning (Reliability Part 3). Taylor & Francis Group, LLC.; 2009

[5] IEEE. Bronze Book on Recommended Practice for Energy Conservation and Cost-Effective Planning in Industrial Facilities. New York: IEEE; 1993

[6] Al-Shaalan A. Reliability/cost tradeoff evaluation for interconnected electric power systems. International Journal of Computing and Digital Systems. 2017;**6**(6):371-374. ISSN 2210-142X

[7] Al-Shaalan A. Fast Fourier Transform (FFT) for Reliability Evaluation of Smart Unit Energy Production. IEEE-Xplore Publications. IEEE Xplore; Date Added to IEEE Xplore; 2018

[8] Al-Shaalan A. Reliability evaluation in generation expansion planning based on the expected energy not supplied. Journal of King Saud University–Engineering Sciences. 2012;**24**(1):11-18

[9] Brown RE. Electric Power Distribution Reliability. 2nd ed. Taylor & Francis Group, LLC; 2009 (Chapter 7)

[10] Billinton R. Reliability Evaluation of Transmission and Distribution Systems. Electricity Supply Association of Australia, Monash University. Centre for Electrical Power Engineering; 1995

[11] Billinton R, Allan RN. Reliability Evaluation of Engineering Systems. Springer Science Business Media, LLC; 1992

[12] Billinton R. Power System Reliability Evaluation. New York: Gordon and Breach Science Publishers; 1970

[13] Ministry of Energy, Industry, and Mineral Resources, the National Renewable Energy Program (NREP) for solar PV projects; 29 January 2019

[14] Saudi Electric Company (SEC). Customer-based reliability measures adopted and used by the SEC, distribution sector standards; 12 September 2010

Chapter 9

Microgrid System Reliability

Razzaqul Ahshan

Abstract

This chapter presents the reliability evaluation of a microgrid system considering the intermittency effect of renewable energy sources such as wind. One of the main objectives of constructing a microgrid system is to ensure reliable power supply to loads in the microgrid. Therefore, it is essential to evaluate the reliability of power generation of the microgrid under various uncertainties. This is due to the stochastically varying wind speed and change in microgrid operational modes which are the major factors to influence the generating capacity of the individual generating unit in the microgrid. Reliability models of various subsystems of a 3-MW wind generation system are developed. The impact of stochastically varying wind speed to generate power by the wind turbine system is accounted in developing sub-system reliability model. A microgrid system reliability (MSR) model is developed by integrating the reliability models of wind turbine systems using the system reliability concept. A Monte Carlo simulation technique is utilized to implement the developed reliability models of wind generation and microgrid systems in a Matlab environment. The investigation reveals that maximizing the use of wind generation systems and storage units increases the reliability of power generation of the proposed microgrid system in different operating modes.

Keywords: reliability, microgrid, distributed generation, wind system, modeling and simulation

1. Introduction

Electricity market deregulation, environmental concerns, technology advancement, and an increased trend for reducing the dependency on fossil fuel are the main causes to integrate distributed generation (DG) units into the distribution power network [1, 2]. Generally, DGs have a diverse generation capacity, availability, and primary energy sources. The increasing demand of adding and utilizing such diverse DGs into the distribution power system brought the concept of microgrid. Microgrid is a flexible combination of loads, DG units, storage systems (either centrally or with each generation individually), and associated power conditioning units operating as a single controllable system that provides power or both power and heat to loads [3]. **Figure 1** shows the generic architecture of a microgrid system.

One of the main objectives of having a microgrid system is to supply reliable power to loads in a microgrid domain. The achievement of such an objective becomes critical when a microgrid system consists of renewable energy sources such as wind and/or solar. In the proposed microgrid system, stochastically varying wind creates unpredictable power variation at the output of the wind turbine system. In addition,

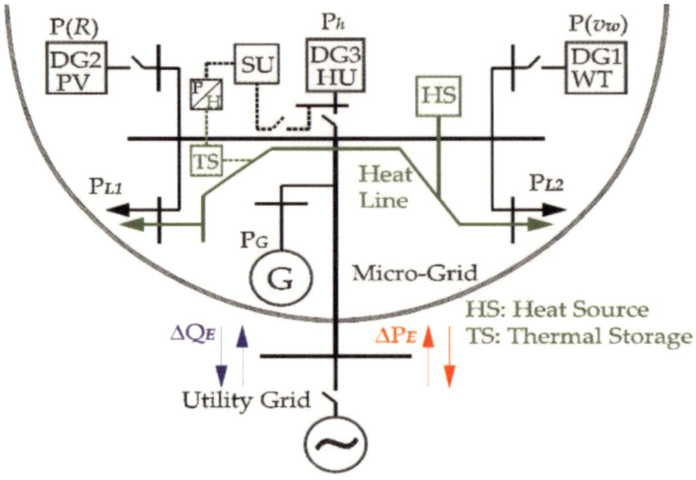

Figure 1.
A generic microgrid system.

such variations in wind speed propagate through all the subsystems in the wind generation system. Therefore, subsystems such as gearbox, generator, and power electronics interfacing units in a wind generation system are also the key factors for producing reliable power by the proposed microgrid system. Thus, it is important to develop the reliability model of the wind generation system including the models of all the subsystems. In addition, consideration of various operation modes of the microgrid system is important to develop a microgrid system reliability model in order to ensure reliable power generation in those operating modes.

The operation, control, and performance characteristics of these microgrids are different because of the contribution of diversity in nature and size of distributed generations in the microgrid. Such diversities of distributed generations include fixed- or variable-speed wind turbines, solar panels, micro-turbines, various types of fuel cells, small hydro, and storage depending upon the sites and resources available. Different control strategies such as load-frequency control, power sharing among parallel converters, central control based on load curve, and active power control are developed for the microgrids presented in [4–15]. The reliability study of a microgrid system is presented in [16], where the concentration is given in a power quality aspect based on the assumption that the microgrid system is a large virtual generator that has the ability to generate sufficient power for loads at various operating conditions. The reliability-based coordination between wind and hydro system is investigated, which shows the adequacy benefits due to the coordination between them when an appropriate number of hydro units are engaged in order to follow the wind speed changes based on the wind power penetration [17].

The reliability and cost assessment of a solar-wind-fuel cell-based microgrid system are investigated in [18]. A recent review study on reliability and economic evaluation of a power system is presented in [19]. It is suggested that the reliability and economic evaluation of power systems with renewable energy sources needs to perform simultaneously. In [20], a new indicator for measuring reliability of a solar-wind microgrid system is showcased. Reliability evaluation of distribution system that consists of wind-storage-photovoltaic system is shown in [21]. It demonstrates the enhancement in reliability of the conventional distribution system using renewable energy sources. In comparison to microgrid architectures and control research, the investigation of the reliability evaluation of microgrid systems has not been much conducted. Therefore, much attention is required to the reliability evaluation of a microgrid system, which primarily showcases a joined combination of renewable energy sources and storage.

Several researchers have studied the reliability assessment of wind turbine generators in power system applications. The application of two-state and multistate models for wind turbine systems is investigated in [22–24]. However, the stochastic variation and interactions of wind speed and thus time-dependent wind power effects are avoided [25]. A Monte Carlo simulation-based method is then used to assess reliability of a wind generation system in [26–29]. All these past studies evaluate reliability of wind turbine systems by determining the available power output using Eq. (1), while the effect of other subsystems such as gearbox, generator, and interfacing power electronics has not been considered:

$$P_o = \begin{cases} 0 & 0 \le v_w \le v_{ciw} \\ (A + Bv_w + Cv_w^2)P_r & v_{ciw} \le v_w \le v_{rw} \\ P_r & v_{rw} \le v_w \le v_{cow} \\ 0 & v_w \ge v_{cow} \end{cases} \tag{1}$$

In Eq. (1), P_o and P_r are rotor output power and rated power of the wind turbine, respectively; v_{ciw}, v_{rw}, and v_{cow} are cut-in, rated, and cut-out wind speed, respectively, whereas the parameters A, B, and C are the functions of cut-in, rated, and cut-out wind speeds.

Moreover, these approaches determine available power only at the output of the WT rotor without considering the role of the other subsystems. In [30], reliability evaluation is carried out only for interfacing power electronics subsystems in order to compare performances of small (1.5 kW) wind generation systems. Furthermore, such reliability assessment of the interfacing power electronics sub-system is performed for a single operating point such as the rated wind speed condition. However, operating conditions of a wind generation system normally vary between cut-in and cut-out wind speed due to the stochastic behavior of the wind speed. Hence, the reliability evaluation of generating power by a wind generation system is important to be performed considering the stochastic variation of wind speed as well as the impact of stochastic wind behavior on different subsystems in a wind generation system. Such considerations are essential in order to achieve better reliability estimation and, thus, to ensure reliable power supply by the microgrid system.

The reliability of power generation by a microgrid system consisting of wind generation, hydro generation, and storage unit is evaluated and presented in this chapter. The microgrid system under study is located at Fermeuse, Newfoundland, Canada. The reliability model of the microgrid system is developed by means of a reliability block diagram. Furthermore, reliability models of the subsystems in conjunction with wind speed data modeling are developed and applied. The use of Monte Carlo simulation in a Matlab environment yields the following outcomes:

a. The proposed microgrid system is able to provide reliable power to an isolated microgrid with a minimum number of wind power generation units (only one) with a reliability of 0.94.

b. However, maximizing the use of wind generation unit (as the number increases) improves the microgrid system reliability to provide dependable power to the isolated microgrid.

c. Due to the lack of sufficient wind, the integration of pumped hydro storage increases the microgrid system reliability to ensure reliable power supply to the isolated microgrid system.

2. Microgrid system reliability

The one-line diagram of the case study's microgrid system shown in **Figure 2** consists of a HGU, a WPGS or a wind farm (WF), and two load areas represented as P_{L1} and P_{L2}. HGU and WPGS are apart from each other by a TL1 (20.12) km transmission line.

Microgrid system reliability (MSR) is a measurement of the system's overall ability to produce and supply electrical power. Such measurement indicates the adequacy of power generation and supply by a microgrid system for a given combination of DG units in the system as well as the subsystems contained in a DG unit. In order to evaluate the reliability of the system shown in **Figure 2**, the combination of DG units and the subsystems contained in a DG unit can be presented by means of a reliability block diagram (RBD) [31] as per **Figure 3**.

Owing to the evaluation of the reliability of the generating power supply by the microgrid system, only DG units are considered. As such, the simplified RBD of the microgrid system is presented in **Figure 4**, wherein all DG units are connected in

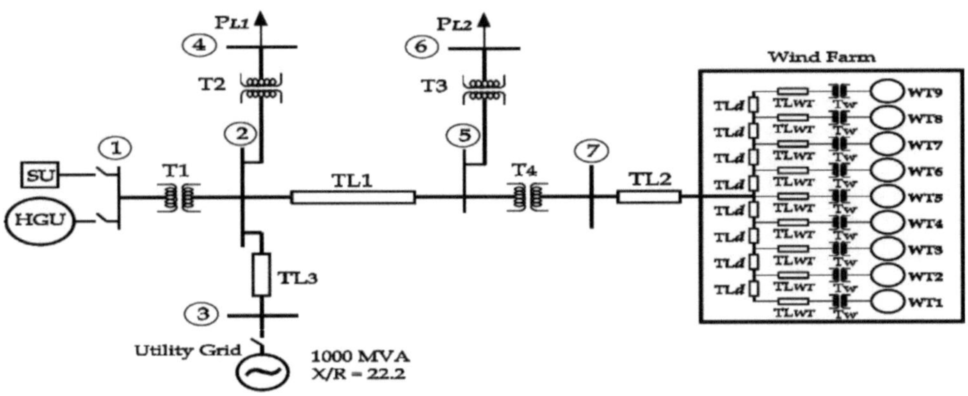

Figure 2.
The single-line diagram of a microgrid system at Fermeuse, Newfoundland, Canada.

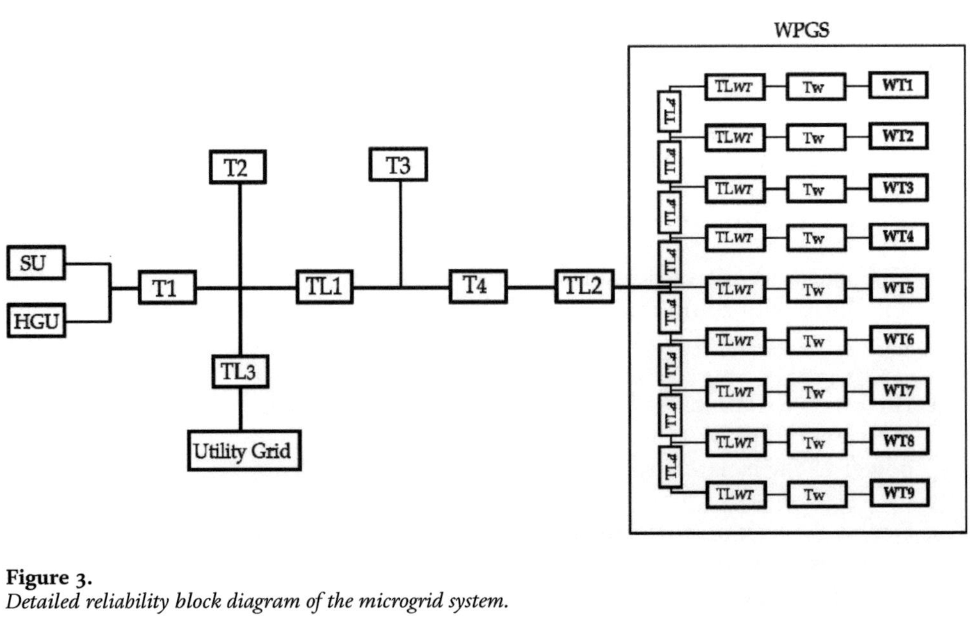

Figure 3.
Detailed reliability block diagram of the microgrid system.

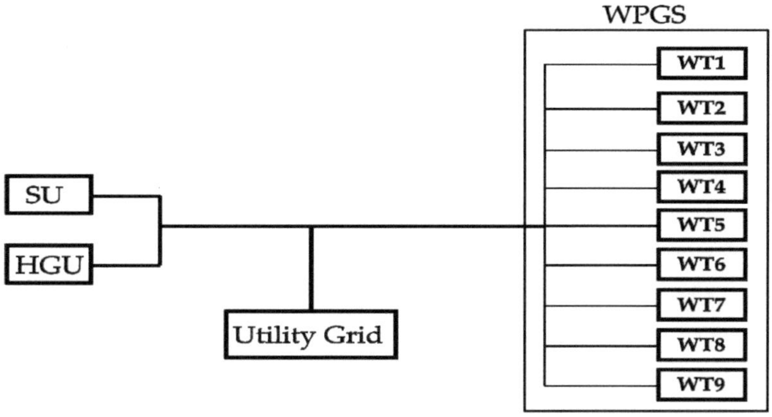

Figure 4.
Simplified reliability block diagram of the microgrid system.

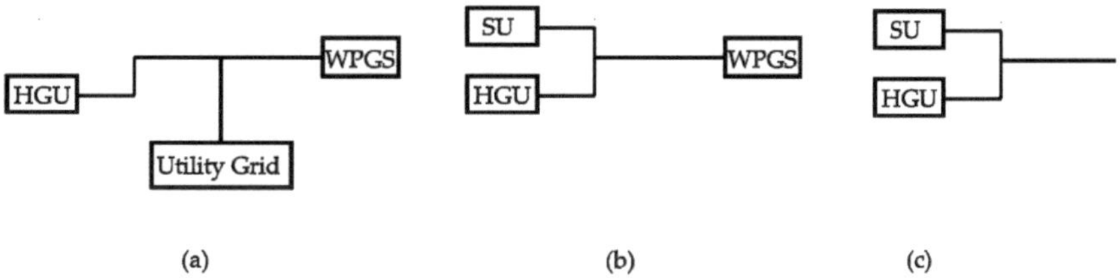

Figure 5.
Reliability block diagram: (a) grid-connected mode, (b) isolated microgrid with wind power generation system, and (c) isolated microgrid without wind power generation system.

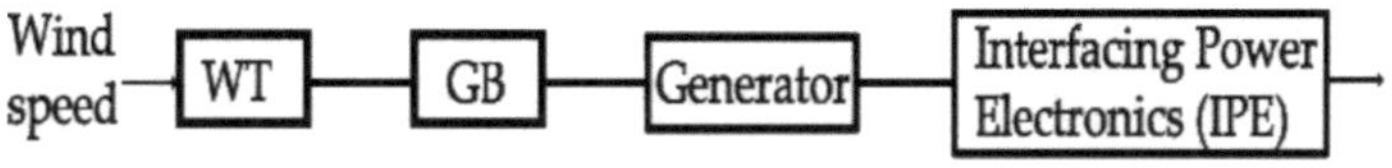

Figure 6.
Reliability block diagram of a wind turbine system.

parallel. However, the RBD of the microgrid system at different operational modes is shown in **Figure 5**.

Moreover, in order to estimate the reliability of a DG unit, its various subsystems may equally be represented by the RBD. The latter is shown in **Figure 6**, which consists of WT or WT rotor, gearbox, generator, and power electronics interfacing circuitry. In this chapter, HGU and utility grid are considered as highly reliable sources of power generation. This is because the HGU at the Fermeuse site produces power at its rated value for an entire year. In addition, the utility grid is also available over the period of a year. The reliability assessment of a storage unit (SU) is beyond the scope of this chapter. However, its reliability is considered based on the fact that the storage system is capable of supplying power to the load during the isolated mode of operation of the microgrid system when wind power generation is unavailable (**Figure 5c**).

3. Reliability modeling

Monte Carlo simulation treats the occurrence of failures as a random event, which mimics the wind speed distribution [32]. For example, in a time series wind speed data, some of the wind speeds are below the cut-in speed of the wind turbine and, as such, will not produce power at the wind turbine output. Such wind speed data can be considered as failure events, which occur randomly. In addition, this research focuses on assessing the reliability of the microgrid system in power generation and supply, considering the wind speed as the primary uncertainty of the system. Hence, Monte Carlo is applied and presented herein.

3.1 Wind speed data modeling

The relation between wind speed and a WT rotor power output is expressed as [33]

$$P_{ro} = 0.5\rho A_{SA} C_p(\lambda, \beta) v^3 \tag{2}$$

where:

- A_{SA} is the swept area covered by the turbine rotor.
- C_p is the power coefficient.
- v_w is the wind velocity.
- β is the pitch angle of rotor blades.
- λ is the tip speed ratio.
- ρ is the air density. Note that for a given WT, A_{SA}, Cp, β, λ, and ρ are constant.

The relation in Eq. (2) can be expressed as

$$P_{ro} \infty v_w^3 \tag{3}$$

Since wind speed is the main factor that creates uncertainty at the power output of a wind energy conversion, wind speed is considered here as the key factor to estimate the MSR. In order to relate the effects of wind speed in calculating the system's overall reliability, wind speed field data modeling is gathered. This is essential because the data itself varies not only from site to site but also according to the hub heights of the wind turbine. Wind speed data modeling for a wind turbine system includes:

a. Identifying best-fit distribution for 1-year wind field data

b. Evaluating the goodness-of-fit test

c. Estimating the distribution parameters

3.1.1 Identification of best-fit distribution

The probability plot method is used to identify the best-fit distribution of the available wind data for a given site and for a given wind turbine hub height. The following steps are taken to accomplish the fitting of the wind data to a distribution:

- Obtain 1-year wind speed data from the site measurement.
- Scale the wind data according to the hub height of the wind turbine using Eq. (4):

$$v_{w2} = v_{w1}\left(\frac{h_2}{h_1}\right)^{\alpha} \tag{4}$$

where h_1 and h_2 are the height of anemometer and hub, respectively; v_{w1} and v_{w2} are the wind velocity at anemometer height and at hub height, respectively; and α is the shear exponent that is expressed as

$$\alpha = (0.096 \log(Z_o) + 0.016 \log(Z_o))^2 + 0.24 \tag{5}$$

where Z_0 is the surface roughness.

- Use Matlab distribution fitting tool to obtain probability plot of the scaled wind data.
- Fit the probability plot of the scaled wind data for different distributions such as normal, log-normal, exponential, and Weibull.
- Identify the distribution corresponding to the best fit of the probability plots.

3.1.2 Goodness-of-fit test

The best-fit distribution of the site wind data is tested for the goodness-of-fit and is performed according to the statistic for MANN'S test:

$$M = \frac{k_1 \sum_{i=k_1+1}^{r-1} \left[\left(\ln\left(v_{w_{i+1}}\right) - \ln\left(v_{w_i}\right) \right) / M_i \right]}{k_2 \sum_{i=1}^{k_1} \left[\left(\ln\left(v_{w_{i+1}}\right) - \ln\left(v_{w_i}\right) \right) / M_i \right]} \tag{6}$$

3.1.3 Distribution parameter estimation

To determine the Weibull distribution parameters, the least-squares technique is used because of its accuracy to fit a straight line in a given data points. In this approach, the wind speed field data are transformed to Weibull distribution to fit to a linear regression line as in Eq. (7):

$$y_i = a + bx_i \tag{7}$$

where

$$x_i = \ln v_{wi} \tag{8}$$

$$y_i = Z_i \tag{9}$$

$$a = -\beta_{ws} \ln \theta \tag{10}$$

$$b = \beta_{ws} \tag{11}$$

The values of a and b are determined from the least-squares fit using Eqs. (8) and (9).

By knowing the values a and b, the Weibull parameters are determined as follows:

$$\theta_{ws} = \exp\left(-\frac{a}{b}\right) \tag{12}$$

$$\beta_{ws} = b \tag{13}$$

where θ_{ws} and β_{ws} are defined as the scale and shape parameters, respectively, for wind speed field data.

3.2 Wind power generation system

According to the microgrid configuration, all nine WTs in WPGS are connected in parallel, which are shown in the simplified RBD in **Figure 4**. In order to estimate the reliability of power generation by the WPGS, a single WT system is considered because all of them are identical both in terms of topology and subsystems context. A WT system comprising of different subsystems is shown in **Figure 6**. The different subsystems are connected in series because failure of power generation by any of the subsystems is considered as a failure of the WT system to generate power. The modeling of the reliability estimation of different subsystems in a WT system is described in the following subsections:

3.2.1 Wind turbine rotor

The wind speed field data model provides information about the shape parameter and scale factor for a Weibull distribution. Such parameters are used to generate a series of random wind speed data that follow a Weibull distribution pattern. Randomly generated data are used to determine power generation by the WT using Eq. (2), which represents a Weibull distribution of power generation. Weibull parameters are determined using the parameter estimation technique described in Section 3.1. These are defined as θ_{tp} and β_{tp}. Thus, the WT's rotor reliability R_{tp} can be expressed as

$$R_{tp} = \exp\left[-\left(\frac{P_{ciw}}{\theta_{tp}}\right)^{\beta_{tp}}\right] - \exp\left[-\left(\frac{P_{cow}}{\theta_{tp}}\right)^{\beta_{tp}}\right] \tag{14}$$

where θ_{tp} and β_{tp} are defined as shape parameter and scale factor for power distribution. P_{ciw} and P_{cow} are the power at cut-in and cut-out wind speed, respectively.

The reliability of generating power at the *ith* wind speed, R_{Pi}, can be expressed as

$$R_{Pi} = \exp\left[-\left(\frac{P_i}{\theta_{tp}}\right)^{\beta_{tp}}\right] \tag{15}$$

where P_i is the power for *ith* wind speed in between cut-in and cut-out regions.

3.2.2 Gearbox

Weibull parameters obtained from field data modeling are utilized to produce a set of random wind data. Such data are used to determine the wind turbine speed using Eq. (16):

$$\omega_{wt} = \frac{\lambda v_w}{R_t} \tag{16}$$

where ω_{wt} is the wind turbine speed and R_t is the turbine radius, respectively. The wind turbine speed is also the speed seen by the gearbox's low-speed shaft. This can be represented as a Weibull distribution of speed. Such a distribution is utilized to estimate the shape parameter and the scale factor of the gearbox. Its reliability R_{gb} can be expressed as

$$R_{gb} = \exp\left[-\left(\frac{\omega_{wt,s}}{\theta_{gb}}\right)^{\beta_{gb}}\right] - \exp\left[-\left(\frac{\omega_{wt,m}}{\theta_{gb}}\right)^{\beta_{gb}}\right] \tag{17}$$

where:

- $\omega_{wt,s}$ is the starting speed of the wind turbine.
- θ_{gb} and β_{gb} are the shape parameter and scale factor for speed seen by the gearbox.
- $\omega_{wt,m}$ is the maximum operating speed of the wind turbine.

The reliability at the *ith* speed seen by the gearbox, $R_{gb,wti}$, can be estimated as

$$R_{gb_{wt,i}} = \exp\left[-\left(\frac{\omega_{wt,i}}{\theta_{gb}}\right)^{\beta_{gb}}\right] \tag{18}$$

where $\omega_{wt,I}$ is the *ith* speed of the WT seen by the gearbox.

3.2.3 Generator

In order to account the effect of wind speed in estimating the wind generator's reliability of generating power, the estimation of Weibull parameters by using field data is shown herein. Such parameters are utilized to generate a set of random wind speed data. Power generated by the WT is then determined using Eq. (2). However, the power at the generator output depends on the gearbox efficiency and various losses in the generator. Efficiency of the gearbox (0.95) and generator (0.95) is considered as 90%, which is observed from the system modeling and simulation. The power at the generator output can be determined as 90% of the power at the turbine output. Thus, a power distribution at the generator output can be obtained, which also follows a Weibull distribution. This, in turn, is used to estimate Weibull distribution parameters using the least-squares parameter estimation technique. After knowing the distribution parameters of the generator output power, the reliability of generating power by the generator, R_g, can be evaluated as

$$R_g = \exp\left[-\left(\frac{P_{g,ciw}}{\theta_{gp}}\right)^{\beta_{gp}}\right] - \exp\left[-\left(\frac{P_{g,cow}}{\theta_{gp}}\right)^{\beta_{gp}}\right] \tag{19}$$

where:

- θ_{gp} and β_{gp} are considered as shape parameter and scale factor for the generator power distribution.
- $P_{g,ciw}$ and $P_{g,cow}$ are the generator power at the cut-in and cut-out wind speeds, respectively.

The reliability of generating power $P_{g,i}$ of the generator, $RP_{g,i}$, can be expressed as

$$R_{P_{g,i}} = \exp\left[-\left(\frac{P_{g,i}}{\theta_{gp}}\right)^{\beta_{gp}}\right] \tag{20}$$

where $P_{g,I}$ is the generator power at the *ith* wind speed in between cut-in and cut-out regions.

3.2.4 Power electronics interfacing system

An interfacing power electronics (IPE) system in a doubly fed induction generator-based WT consists of a back-to-back pulse width modulated (PWM) converter as shown in **Figure 7**. The components in the IPE system are diodes, IGBT switches, and a DC bus capacitor. The reliability model of such a system can be developed based on the relationship between the lifetime and failure rate of the components in the system. These are determined considering the junction temperature as a covariate. The junction temperature, T_j, of a semiconductor device can be calculated as [34]

$$T_j = T_a + P_l R_{ja} \tag{21}$$

where P_l, T_a, and R_{ja} are the power loss of a component, the ambient temperature, and the junction resistance, respectively. A reliability model of a power conditioning system for a small (1.5 kW) wind energy conversion system is developed by considering power loss only at a rated wind speed operating condition.

However, it is to be noted that power losses in the semiconductor components vary according to the wind speed variation at the wind turbine input. Thus, a power loss variation in the semiconductor component is important to be considered as a stress factor in order to calculate the lifetime of the components instead of using power loss quantity for a single operating condition. Hence, Eq. (21) can be expressed as

$$T_{j_i} = T_a + P_{l_i} R_{ja} \tag{22}$$

where:

- P_{l_i} is the power loss of a component at the ith wind speed.
- T_{j_i} is the component junction temperature at the ith wind speed
- Junction resistance is assumed to be constant for all wind speed.

In an IPE system, there are two types of semiconductor components, namely, diode and IGBT switches. Two types of power losses such as conduction losses and switching losses occur in such components. The conduction loss, $P_{cl,d}$, and the switching loss, $P_{sl,d}$, of a diode can be expressed as [35, 36]

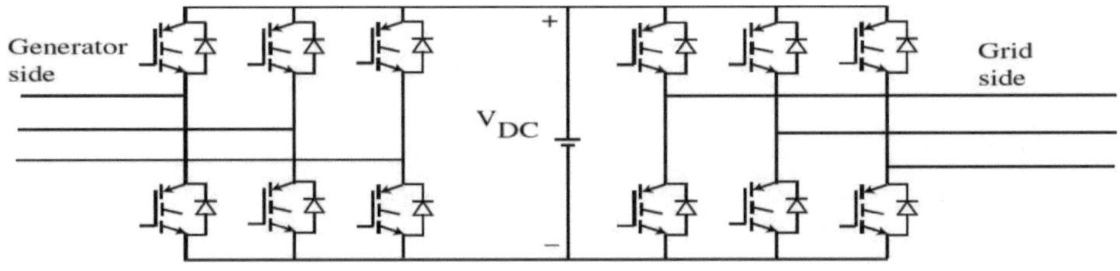

Figure 7.
Interfacing power electronics system of a doubly fed induction generator-based wind turbine system.

$$P_{cl,d} = \left(\frac{1}{8} - \frac{M}{3\pi}\cos\varphi\right)R_d I_{mo}^2 + \left(\frac{1}{2\pi} - \frac{M}{8}\cos\varphi\right)V_{FO} I_{mo} \tag{23}$$

$$P_{sl,d} = \frac{1}{\pi} f_s E_{sr} \frac{V_{dc} I_{mo}}{V_{ref,d} I_{ref,d}} \tag{24}$$

Total power losses of diodes, $P_{tl,d}$, in the IPE system can be expressed as the sum of the conduction loss, $P_{cl,d}$, for the total number of diodes. The switching loss, $P_{sl,d}$, for the total number of switches in the system can be expressed as

$$P_{tl,d} = n\left(\frac{1}{8} - \frac{M}{3\pi}\cos\varphi\right)R_d I_{mo}^2 + n\left(\frac{1}{2\pi} - \frac{M}{8}\cos\varphi\right)V_{FO} I_{mo} + n\frac{1}{\pi} f_s E_{sr} \frac{V_{dc} I_{mo}}{V_{ref,d} I_{ref,d}} \tag{25}$$

where:

- M is the modulation index ($0 \leq M \leq 1$).
- I_{mo} is maximum output current of the inverter.
- n is the number of semiconductor components.
- V_{FO} and R_d are the diode threshold voltage and resistance, respectively.
- f_s is the switching frequency.
- E_{er} is the rated switching loss energy given for the commutation voltage.
- $V_{ref,d}$ and $I_{ref,d}$, V_{dc}, and I_{dc} are the actual commutation voltage and current, respectively.
- φ is the angle between voltage and current.

The conduction loss, $P_{cl,IGBT}$, and switching loss $P_{sl,IGBT}$ of an IGBT switch can be expressed as [37]

$$P_{cl,IGBT} = \left(\frac{1}{8} + \frac{M}{3\pi}\cos\varphi\right)R_{ce} I_{mo}^2 + \left(\frac{1}{2\pi + \frac{M}{8}\cos\varphi}\right)V_{CEO} I_{mo} \tag{26}$$

$$P_{sl,IGBT} = \frac{1}{\pi} f_s (E_{on} + E_{off}) \frac{V_{dc} I_{mo}}{V_{ref,IGBT} I_{ref,IGBT}} \tag{27}$$

Total power losses of switches, $P_{tl,IGBT}$, in the IPE system can be expressed as the sum of the conduction loss, $P_{cl,IGBT}$, for total number of diodes. The switching loss, $P_{sl,IGBT}$, for the total number of switches in the system can be expressed as

$$P_{tl,IGBT} = n\left(\frac{1}{8} + \frac{M}{3\pi}\cos\varphi\right)R_{ce} I_{mo}^2 + n\left(\frac{1}{2\pi} + \frac{M}{8}\cos\varphi\right)V_{CEO} I_{mo} + n\frac{1}{\pi} f_s (E_{on} + E_{off}) \frac{V_{dc} I_{mo}}{V_{ref,IGBT} I_{ref,IGBT}} \tag{28}$$

where:

- V_{CEO} and R_{ce} are the IGBT threshold voltage and on-state resistance, respectively.

- The reference commutation voltage and current are $V_{ref,IGBT}$ and $I_{ref,IGBT}$.
- V_{dc} is the actual commutation voltage.
- E_{on} and E_{off} are the turn-on and turn-off energies of IGBT.

The lifetime, $L(T_{ji})$, of a component for *ith* wind speed can be expressed as

$$L(T_{ji}) = L_o \exp\left(-B\Delta T_{ji}\right) \tag{29}$$

where:

- L_o is the quantitative normal life measurement (assumed to be 10^6).
- $B = \frac{E_A}{K}$, where K is the Boltzmann's constant (=8.6×10^{-5} eV/K) and E_A is the activation energy (= 0.2 eV) for typical semiconductor components.
- ΔT_{ji} is the variation in junction temperature for the *ith* wind speed and can be expressed as

$$\Delta T_{ji} = \frac{1}{T_a} - \frac{1}{T_{ji}} \tag{30}$$

The failure rate of a component for *ith* wind speed can be defined as

$$\tau_i = \frac{1}{L(T_{ji})} \tag{31}$$

Using Eq. (31), a distribution of failure rates for a set of wind speed data for a semiconductor component can be generated. The components in the IPE system are considered as an in-series connection from the reliability point of view, because the IPE system fails, if any one of the components breaks down in the IPE system. Thus, the failure rates for different components are added to determine the failure rate of the IPE system for the *ith* wind speed. Hence, a distribution of failure rates for the IPE system can be generated for a series of wind speed data.

A least-squares technique is then used to determine the distribution parameters. By knowing the distribution parameters, the reliability of the IPE system, R_{IPE}, can be modeled as

$$R_{\text{IPE}} = \exp\left[-\left(\frac{\tau_{ciw}}{\theta_{\text{IPE}}}\right)^{\beta_{\text{IPE}}}\right] - \exp\left[\left(\frac{\tau_{cow}}{\theta_{\text{IPE}}}\right)^{\beta_{\text{IPE}}}\right] \tag{32}$$

Hence:

- θ_{IPE} and β_{IPE} are defined as the shape parameter and the scale factor for the failure rate distribution of the IPE system.
- τ_{ciw} and τ_{cow} are failure rates of IPE system at cut-in and cut-out wind speeds, respectively.

The reliability of a component in IPE system, R_{IPEC}, can be expressed as

$$R_{\text{IPE}_C} = \exp\left[-\left(\frac{\tau_{ciw_C}}{\theta_{\text{IPE}_C}}\right)^{\beta_{\text{IPE}_C}}\right] - \exp\left[\left(\frac{\tau_{cow_C}}{\theta_{\text{IPE}_C}}\right)^{\beta_{\text{IPE}_C}}\right] \tag{33}$$

where:

- θ_{IPE_C} and β_{IPE_C} are defined as the shape parameter and the scale factor for the failure rate distribution of a component.

- τ_{ciw_C} and τ_{cow_C} are failure rates at cut-in and cut-out wind speeds for a component, respectively.

The reliability of a WT system, R_{wts}, can now be expressed as

$$R_{wts} = R_{tp} \times R_{gb} \times R_g \times R_{IPE} \tag{34}$$

In WPGS, all nine WTs are connected in parallel with identical configuration. Hence, the reliability of the WPGS, R_{WPGS}, can be expressed as

$$R_{WPGS} = \left[1 - (1 - R_{wts})^N\right] \tag{35}$$

where N is the number of WT system in a WPGS.

3.3 Microgrid reliability model

Figure 4 shows the simplified RBD of the microgrid system, where all DG units are connected in parallel. In addition, SU is considered as a power-generating unit since it will supply power to the load during an isolated mode of operation of the microgrid. Assuming the reliability of the HGU as R_{HGU} and utility grid as R_{UG}, the overall microgrid system reliability, R_{MSR}, can be modeled as

$$R_{MSR} = \left[1 - (1 - R_{wts})^N (1 - R_{HGU})(1 - R_{UG})\right] \tag{36}$$

However, the microgrid system operates in three different modes, which are shown in **Figure 5**. The MSR can also be modeled according to their operating modes. **Figure 5a** shows the grid-connected mode of operation where all DG or generation units are connected with the utility grid. Thus, the MSR pertaining to the grid-connected mode of operation, $R_{MSR_{M1}}$, can be expressed by the similar model presented in Eq. (36). Therefore,

$$R_{MSR_{M1}} = \left[1 - (1 - R_{wts})^N (1 - R_{HGU})(1 - R_{UG})\right] \tag{37}$$

Figure 5b represents an isolated microgrid system with WPGS. In addition, the storage unit is not working as a generation unit in this mode of operation. Thus, the MSR during isolated operation with WPGS, $R_{MSR_{M2}}$, can be defined as

$$R_{MSR_{M2}} = \left[1 - (1 - R_{wts})^N (1 - R_{HGU})\right] \tag{38}$$

Furthermore, **Figure 5c** shows an isolated microgrid without WPGS mode where the SU operates as a generation unit. Assuming that the reliability of the SU is R_{SU}, hence, the MSR during this mode, $R_{MSR_{M3}}$, can be written as

$$R_{MSR_{M3}} = [1 - (1 - R_{HGU})(1 - R_{HGU})] \tag{39}$$

4. Implementation of the microgrid model

In order to implement the developed MSR model so as to evaluate the power generation reliability of the proposed microgrid system, Monte Carlo simulation is performed using Matlab. The flow diagram is shown in **Figure 8** and is explained in steps 1–5. The model of the MSR and the reliability evaluation of various operating modes of the proposed microgrid are implemented using Matlab code according to the flow chart shown in **Figure 9** and explained in steps 6–7.

Step 1: Wind speed field data model

- Field data collection and distribution identification using probability plots
- Goodness-of-fit test for selecting the distribution of wind speed
- Calculating the distribution parameter using Eqs. (12) and (13)
- Generating a series of random data as the input for the next steps of the reliability flow diagram

Step 2: Reliability of power generation by WT rotor

- WT rotor output power distribution generation

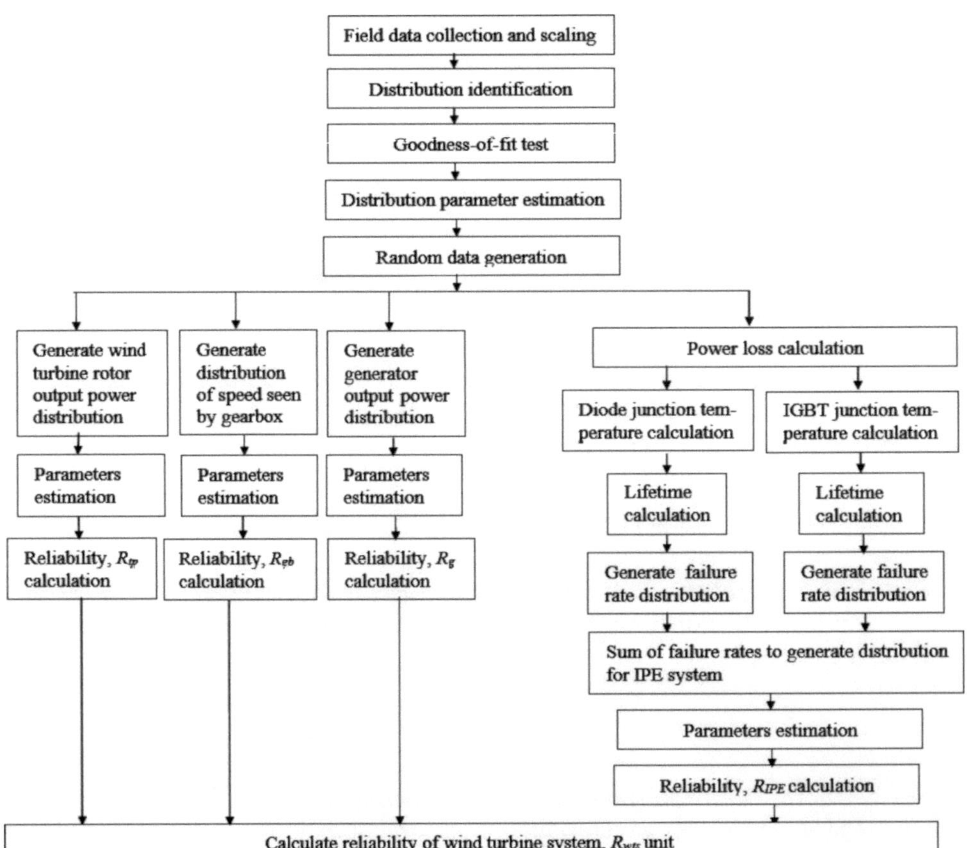

Figure 8.
Flow diagram for reliability calculation of wind generation subsystems.

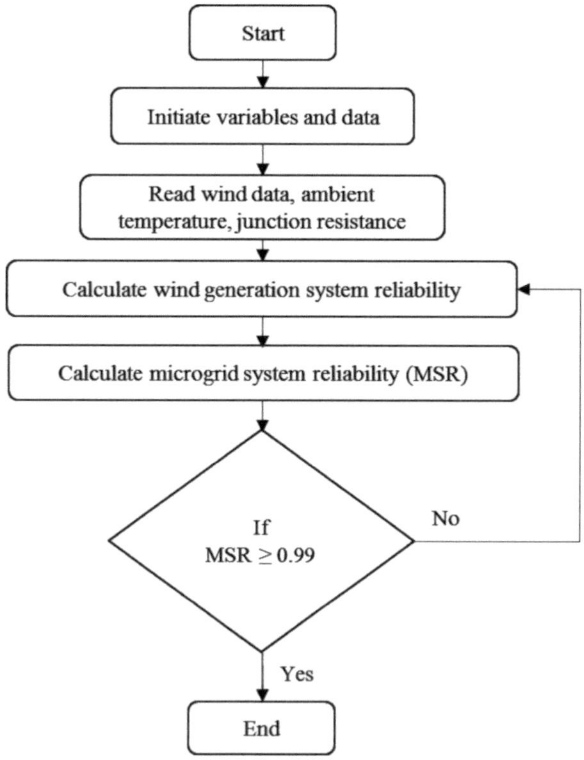

Figure 9.
Flow chart for calculating the microgrid system reliability.

- Parameter estimation for WT rotor power distribution
- Reliability calculation using Eq. (14)

Step 3: Reliability of gearbox

- Determining speed distribution seen by the gearbox
- Speed distribution parameter calculation using least-squares technique
- Reliability calculation using Eq. (17)

Step 4: Reliability of generator

- Generator output power distribution generation
- Distribution parameter determination using least-squares technique
- Reliability evaluation of generator output power using Eq. (19)

Step 5: Reliability of interfacing power electronics

- Power loss calculation of diodes and IGBTs in the IPE system using Eqs. (25) and (28)
- Failure rate distribution generation for diodes and IGBT switches

- Estimating parameter of failure rate distribution of IPE system
- Calculating reliability using Eq. (32)

Step 6: Reliability of DG units

- Reliability calculation of a WT system using Eq. (34)
- Determining reliability of WPGS using Eq. (35)
- Assuming reliability for HGU and SU

Step 7: Reliability of microgrid system

- MSR calculation using Eqs. (36)–(39) for various operational modes

5. Simulation results

The reliability model and its implementation procedure described in the preceding sections are performed to determine probability distribution parameters as well as the reliability of the various subsystems in the wind generation system for stochastically varying wind speed condition. Such reliability estimation is then utilized to determine MSR in various operating modes of the microgrid. The power generation wind speed region of the selected turbine is v_{ciw} = 4 m/s and v_{cow} = 25 m/s. The reliability of HGU and utility grid are selected as 85%, since they are regarded as highly reliable power generation sources. The reliability of the storage unit is assumed to be the same as the IPE system (=0.8144), because these are commonly interfaced through power electronics inverter systems. One-year wind speed data is used for the field data modeling process. Assume that three WT systems can be connected to the isolated microgrid system due to the stability issue.

Figure 10 shows the hourly wind speed field data collected over a 1-year period. Such data is utilized to identify the distribution using probability plot techniques. The probability plots of wind speed field data are shown in **Figure 11**, revealing that the probability of wind speed follows Weibull and Rayleigh distributions closely. However, the Weibull distribution follows the probability of wind speed closer than the Rayleigh distribution. Thus, the Weibull distribution is identified as the best-fit distribution for wind speed data in this study. In order to select Weibull distribution, a goodness-of-fit test is also carried out, and the probability density function of Weibull distribution is shown in **Figure 12**.

A least-squares method is performed to estimate the Weibull distribution parameter, which is shown in **Figure 13**. The shape parameter for wind speed

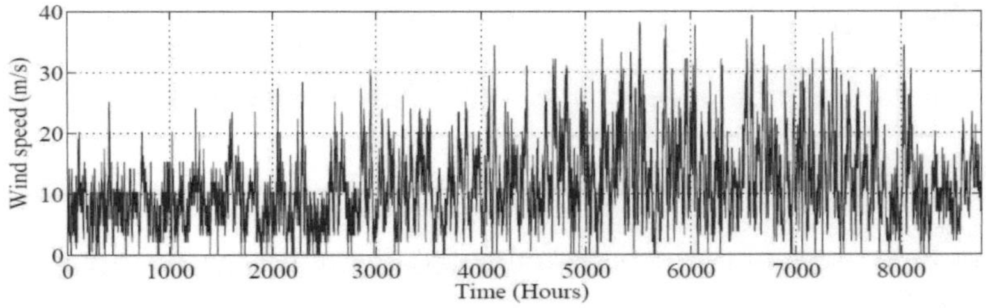

Figure 10.
Wind speed field data.

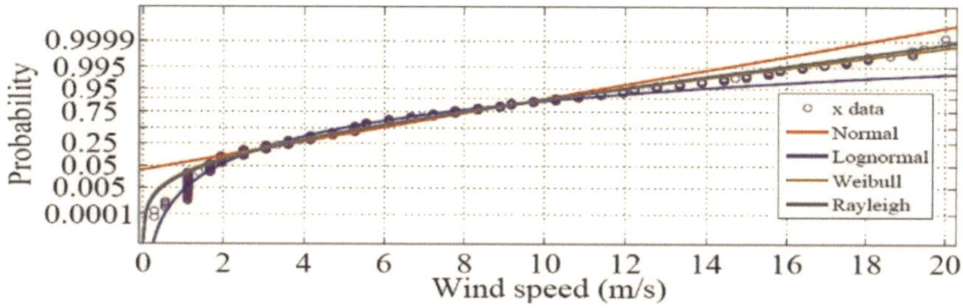

Figure 11.
Probability plots for distribution identification.

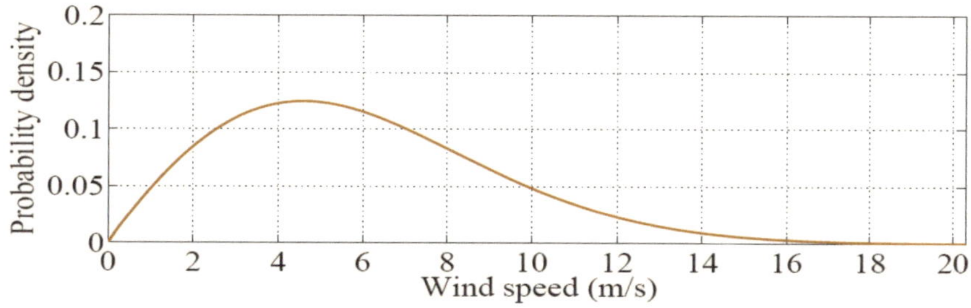

Figure 12.
Probability density function of wind speed data.

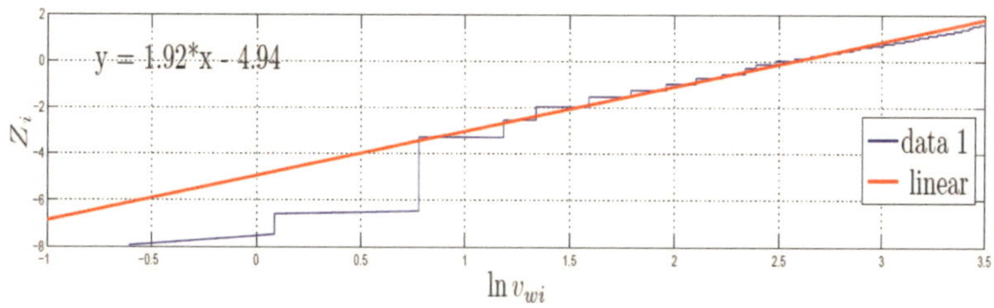

Figure 13.
Least-squares plot for parameter estimation.

β_{ws} = 1.92, and the scale parameter θ_{ws} = 13.1. These parameters are used to generate random wind speed data for reliability evaluation of different subsystems in a wind turbine system.

The results of the reliability calculation for different subsystems in a wind generation system are presented in **Table 1**. The outcomes reveal that the resulting reliability of the wind turbine rotor is 0.9068, while the reliability of gearbox and generator are 0.9107 and 0.9266, respectively. However, the reliability of generating power for the IPE sub-system is only 0.8144. These findings indicate that the IPE sub-system in a variable-speed wind generator system is less reliable than the other subsystems. **Table 2** presents the reliability results of DG units such as WT system, WPGS, HGU, SU, and utility grid. The reliability of a WT system and a WPGS is calculated based on the model derived in this study; however, the reliability of HGU, SU, and utility grid is assumed based on their availability in operation. The overall reliability of a wind turbine system is 0.6232. Since nine WT systems are connected in parallel in the WPGS, the calculated reliability of WPGS is significantly high.

The reliability estimation results of the microgrid system during various operational modes are presented in **Table 3**. The MSR during grid-connected mode is

Sub-systems	Distribution parameters		Sub-systems parameters		Reliability
WT rotor	θ_{tp}	β_{tp}	P_{ciw}	P_{cow}	R_{tp}
	1560.58	1.422	77	3000	0.9068
Gearbox	θ_{gb}	β_{gb}	$\omega_{wt,s}$	$\omega_{wt,\mathrm{m}}$	R_{gb}
	13.73	3.33	4.1	18.4	0.9107
Generator	θ_{g}	β_{g}	$P_{g,ciw}$	$P_{g,cow}$	R_{g}
	1354	1.4142	73	2850	0.9266
IPE system	θ_{IPE}	β_{IPE}	τ_{ciw}	τ_{cow}	R_{IPE}
	1.158	2.658e-5	0.0202e-4	0.4821e-4	0.8144

Table 1.
Reliability results of different subsystems in a variable-speed wind generator system.

DG units	Reliability	DG units	Reliability
WT system	R_{wts}	HGU	R_{HGU}
	0.6232		0.85
WPGS	R_{WPGS}	SU	R_{SU}
	0.9998		0.8144

Table 2.
Reliability results of distributed generation units.

Microgrid operational modes	Reliability
Grid-connected mode	$R_{MSR_{M1}}$
	0.9999
Isolated microgrid with WPGS: number of WTs in WPGS (1, 2, 3, 4)	$R_{MSR_{M2}}$
	0.94, 0.97, 0.99, 0.997
Isolated microgrid without WPGS	$R_{MSR_{M3}}$
	0.99

Table 3.
Reliability results of microgrid system.

higher than the other operational modes because all DG units are operating during this mode. Moreover, this mode has two generation sources which are assumed as highly reliable in power generation and supply.

On the other hand, MSR during isolated microgrid with WPGS varies depending on the number of WT system operating in the WPGS. It is worth mentioning that in an isolated microgrid system, all WTs in WPGS do not operate due to stability issues. This issue will occur since all WTs in WPGS will require reactive power for their operation during isolated mode. However, in an isolated mode, there is no such reactive power generation source to provide sufficient reactive power for all nine WT systems. Thus, the reliability calculation is carried out for a different number of WT systems in the WPGS, and the various reliability indices are found. On the other hand, it is important to note that the minimum reliability index found is 0.94, which is high.

Moreover, the reliability level during this mode of operation (**Figure 5b**) can also be increased by adding more generation sources within the maximum number of constraint (maximum number of WT system).

The reliability of the microgrid system without WPGS is calculated as 0.97, which is higher than that of a microgrid system with WPGS. This is due to the combination of generation sources in this mode of operation (**Figure 5c**), which are highly reliable than the generation source (such as WT) in the WPGS. The results of the reliability evaluation shows that the proposed microgrid system has the significant ability to generate sufficient power to ensure the reliable power supply in all operating modes. The reliability indices found in this study reveal that a microgrid system consisting of renewable energy sources such as wind, hydro, and storage is reliable in generating and supplying power.

6. Conclusions

This chapter discussed the reliability assessment of a microgrid system, comprising variable-speed wind generator units. This research was carried out on a microgrid system located at Fermeuse, Newfoundland, Canada. The mathematical model of microgrid system reliability is developed based on the reliability block diagram (RBD) concept. In addition, the reliability model of various subsystems in a variable-speed wind generator unit is developed considering the impact of stochastically varying wind speed. The developed microgrid system reliability model is implemented through Monte Carlo simulation using Matlab coding. The obtained results are presented and discussed.

- The reliability performance of generating and supplying reliable power by the case study microgrid system during its various operational modes is found to equal 0.99 (grid-connected mode), 0.99 (isolated microgrid with WPGS), and 0.99 (isolated microgrid without WPGS).

- This suggests that the microgrid has the ability to generate and supply power to the loads in a microgrid domain with a high degree of reliability. Such a reliability level is achieved due to maximizing the use of renewable power. The latter stems from wind generation systems as well as storage units.

- It is the authors' view that this reliability evaluation approach may be applied to assess the reliability of microgrid systems containing other intermittent energy sources such as solar.

The developed and presented method in this chapter is implemented using simulation. However, this method is neither implemented in real time, nor is it sold to industry yet. This method needs further investigation to include other renewable sources such as solar-based ones. In addition, an experimental investigation is also required, which in turn may prove challenging, as a number of key issues need to be addressed.

At present, the author is further researching the possibility of applying described method for a microgrid that consists of a solar photovoltaic system and may be applicable to hot weather conditions.

Acknowledgements

This work is supported by a research grant from the National Science and Engineering Research Council (NSERC) of Canada, the Atlantic Innovation Fund

(AIF) of Canada, and Memorial University of Newfoundland. The author also would like to acknowledge the utility company, Newfoundland Power, Canada, for providing the system information and data.

Conflict of interest

There is no conflict of interest.

Abbreviations

DC	direct current
DG	distributed generation
GB	gearbox
HGU	hydro generation unit
IGBT	insulated-gate bipolar transistor
IPE	interfacing power electronics
MSR	microgrid system reliability
PWM	pulse width modulation
RBD	reliability block diagram
SU	storage unit
WF	wind farm
WPGS	wind power generation system
WT	wind turbine

Author details

Razzaqul Ahshan
Department of Electrical and Computer Engineering, Sultan Qaboos University, Muscat, Sultanate of Oman

*Address all correspondence to: razzaqul@squ.edu.om

References

[1] Ahshan R, Iqbal MT, Mann GKI, Quaicoe JE. Modeling and analysis of a micro-grid system powered by renewable energy sources. The Open Renewable Energy Journal. 2013;**6**:7-22

[2] Ahshan R, Shafiq M, Hosseinzadeh N, Al-Badi A. Distributed wind systems for moderate wind speed sites. In: 5th Int. Conf. on Renewable Energy Generation and Applications; UAE: Al-Ain; 2018

[3] Lasseter RH. Microgrids. In: Proc. of IEEE Power Engineering Society Winter Meeting; 2002. pp. 1-4

[4] Barnes M, Dimeas A, Engler A, Fitzer C, Hatziargyriou N, Jones C, et al. Microgrid laboratory facilities. In: IEEE International Conference on Future Power System; 2005

[5] Lopes JAP, Madureira AG, Moreira CCLM. A view of microgrids. WIREs Energy and Environment. 2013;**2**: 86-103. DOI: 10.1002/wene.34

[6] Hatziargyriou N, Asano H, Iravani MR, Marnay C. Microgrids—An overview of ongoing research, development and demonstration projects. In: IEEE Power and Energy Magazine, LBNL-62937; 2007. pp. 78-94

[7] Morozumi S. Overview of microgrid research and development activities in Japan. In: IEEE Symposium on Micro-grids; 2006

[8] Katiraei F, Iravani MR, Lehn PW. Small signal dynamic model of a microgrid including conventional and electronically interfaced distributed resources. IET Generation Transmission and Distribution. 2007;**1**(3):369-378

[9] Katiraei F, Abbey C, Bahry R. Analysis of voltage regulation problem for 25kV distribution network with distributed generation. In: Proc. of IEEE Power Engineering Society General Meeting; 2006. pp. 1-8

[10] Ibrahim H, Ghandour M, Dimitrova M, Ilinca A, Perron J. Integration of wind energy into electricity systems: Technical challenges and actual solutions. Energy Procedia. 2011;**6**: 815-824

[11] Shahabi M, Haghifam MR, Mohamadian M, Nabavi-Niaki SA. Microgrid dynamic performance improvement using a doubly fed induction wind generator. IEEE Transactions on Energy Conversion. 2009;**24**(1):137-145

[12] Majumder R, Ghosh A, Ledwich G, Zare F. Load sharing and power quality enhanced operation of a distributed micro-grid. IET Renewable Power Generation. 2009;**3**(2):109-119

[13] Nayar C. Innovative remote microgrid systems. International Journal of Environment and Sustainability. 2012;**1**(3):53-65

[14] Kawasaki K, Matsumura S, Iwabu K, Fujimuram N, Iima T. Autonomous dispersed control system for independent microgrid. Journal of Electrical Engineering, Japan. 2009; **166**(1):1121-1127

[15] Li X, Song Y, Han S. Frequency control in microgrid power system combined with electrolyzer system and fuzzy PI controller. Journal of Power Sources. 2008;**180**:468-475

[16] Basu AK, Chaowdhury SP, Chaowdhury S, Ray D, Crossley PA. Reliability study of a microgrid system. IEEE Transactions on Power Systems. 2006;**21**(4):1821-1831

[17] Karki R, Hu P, Billinton R. Reliability evaluation considering wind

and hydropower coordination. IEEE Transactions on Power Systems. 2010; **25**(2):685-603

[18] Tanrioven M. Reliability and cost-benefits of adding alternate power sources to an independent micro-grid community. Journal of Power Sources. 2005;**150**:136-149

[19] Zhou P, Jin RY, Fan LW. Reliability and economic evaluation of power system with renewables: A review. Renewable and Sustainable Energy Reviews. 2016;**58**:537-547

[20] Acuna L, Padilla RV, Mercado AS. Measuring reliability of hybrid photovoltaic-wind energy systems: A new indicator. Renewable Energy. 2017; **106**:68-77

[21] Adefarati T, Bansal RC. Reliability assessment of distribution system with the integration of renewable distributed generation. Applied Energy. 2017;**185**: 158-171

[22] Li J, Wei W. Probabilistic evaluation of available power of a renewable generation system consisting of wind turbine and storage batteries: A Markov chain method. Journal of Renewable and Sustainable Energy. 2014;**6**(1): 1493-1501

[23] Liu X, Chowdhury AA, Koval DO. Reliability evaluation of a wind-diesel-battery hybrid power system. In: IEEE Industrial and Commercial Power Systems Technical Conference; 2008

[24] Wang L, Singh C. Adequacy assessment of power-generating systems including wind power integration based on ant colony system algorithm. In: IEEE Lausanne Power Tech; 2007

[25] Wen J, Zheng Y, Donghan F. A review on reliability assessment for wind power. Renewable and Sustainable Energy Reviews. 2009;**13**:2485-2494

[26] Bhuiyan FA, Yazdani A. Reliability assessment of a wind power system with integrated energy storage. IET Renewable Power Generation. 2010;**4**(3):211-220

[27] Choi J, Park JM, Shahidehpour M. Probabilistic reliability evaluation of composite power systems including wind turbine generators. In: Proc. of IEEE International Conference on Probabilistic Method Applied to Power Systems; 2010. pp. 802-807

[28] Vallee F, Lobry J, Deblecker O. System reliability assessment method for wind power integration. IEEE Transactions on Power Systems. 2008; **23**(3):1288-1297

[29] Silva AMLD, Manso LAF. Application of Monte-Carlo simulation to generating system well-being analysis considering renewable sources. European Transactions on Electrical Power. 2007;**17**:387-400

[30] Arifujjaman M. Performance and reliability comparison of grid connected small wind turbine systems. PhD dissertation. NL, CA: Department of Electrical and Computer Engineering, Memorial University of Newfoundland, St. John's; 2010

[31] Ebeling CE. An Introduction to Reliability and Maintainability Engineering. USA: Waveland Press; 2010

[32] Vittal S, Teboul M. Performance and reliability analysis of wind turbines using Monte-Carlo methods based on System Transport Theory. In: Proc. of Structural Dynamics and Materials Conference; 2005. pp. 1-8

[33] Oettinger FF, Blackburn DL, Rubin S. Thermal characteristics of power transistors. IEEE Transactions on Reliability. 1976;**23**(8):831-838

[34] Ali MH. Wind Energy Systems: Solutions for Power Quality and Stabilization. USA: CRC Press; 2012

[35] Feix G, Dieckerhoff S, Allmeling J, Schonberger J. Simple methods to calculate IGBT and diode conduction and switching losses. In: Proc. of 13th European Conference on Power Electronics and Applications; 2009. 8-10

[36] Bierhoff MH, Fuchs FW. Semiconductor losses in voltage source and current source IGBT converters based on analytical derivation. In: Proc. of IEEE Power Electrics Specialist Conference; 2004. pp. 2836-2842

[37] Mestha LK, Evans PD. Analysis of on-state losses in PWM inverter. IEE Proceedings. 1989;**136**(4):1989